Springer Topics in Signal Processing

Volume 4

Series Editors

J. Benesty, Montreal, QC, Canada
W. Kellermann, Erlangen, Germany

Springer Topics in Signal Processing

Edited by J. Benesty and W. Kellermann

Vol. 1: Benesty, J.; Chen, J.; Huang, Y.
Microphone Array Signal Processing
250 p. 2008 [978-3-540-78611-5]

Vol. 2: Benesty, J.; Chen, J.; Huang, Y.; Cohen, I.
Noise Reduction in Speech Processing
240 p. 2009 [978-3-642-00295-3]

Vol. 3: Cohen, I.; Benesty, J.; Gannot, S. (Eds.)
Speech Processing in Modern Communication
360 p. 2010 [978-3-642-11129-7]

Vol. 4: Benesty, J.; Paleologu, C.; Gänsler, T.; Ciochină, S.
A Perspective on Stereophonic Acoustic Echo Cancellation
139 p. 2011 [978-3-642-22573-4]

Jacob Benesty · Constantin Paleologu
Tomas Gänsler · Silviu Ciochină

A Perspective on Stereophonic Acoustic Echo Cancellation

Springer

Prof. Dr. Jacob Benesty
INRS-EMT
University of Quebec
H5A 1K6 Montreal, QC
Canada
Email: benesty@emt.inrs.ca

Dr. Tomas Gänsler
mh acoustics LLC
Summit, New Jersey
USA
Email: tfg@mhacoustics.com

Prof. Dr. Constantin Paleologu
University Politehnica of Bucharest
Telecommunications Department
061071 Bucharest
Romania
Email: pale@comm.pub.ro

Prof. Dr. Silviu Ciochină
Politehnica University of Bucharest
Telecommunications Department
061071 Bucharest
Romania
Email: silviu@comm.pub.ro

ISBN 978-3-642-22573-4 ISBN 978-3-642-22574-1 (eBook)

DOI 10.1007/978-3-642-22574-1

Springer Topics in Signal Processing ISSN 1866-2609

e-ISSN 1866-2617

Library of Congress Control Number: 2011933378

Cover Design: WMXDesign GmbH, Heidelberg

Printed on acid-free paper

9 8 7 6 5 4 3 2 1

springer.com

Contents

Chapter 1
Introduction

1.1 Stereophonic Acoustic Echo Cancellation (SAEC)

Research and development of stereophonic echo control systems has been a subject of interest over the last 20+ years. In fact, one of the first papers describing the characteristics of stereophonic echo cancellation was presented in 1991 [1]. In this paper, it is pointed out that the loudspeaker (input) signals are linearly related through non-invertible acoustic room responses (in the case of one source, but not necessarily two or more). The consequence of this linear relationship is that the underlying normal (or Wiener-Hopf) equations to be solved by the adaptive algorithm is an ill-conditioned, or in the worst case, a singular problem. In the singular case, the adaptive filter can drift between candidates in the set of available nonunique solutions, all minimizing the variance of the output error. However, these solutions do not necessarily minimize filter misalignment. As a result, some unstable behavior for certain time varying events may be expected. Even though the problem of nonuniqueness was described, analyzed, and solutions presented in early publications, e.g., [2], [3], [4], many following proposals have been confused over what has to be done to solve the problem correctly. Fundamentally, the core solution to the stereophonic acoustic echo cancellation (SAEC) problem must tackle two issues: (a) provide a proper solution to the inherent ill-posed problem of stereophonic echo cancellation and (b) mitigate the effect that the ill-posed problem has on the convergence rate and tracking of the adaptive algorithm. The former problem (a), can only be solved by manipulating the signals actually transmitted to the near-end (receiving) room, i.e., using a preprocessor on the loudspeaker signals to decorrelate them (or more accurately to reduce the coherence) before the SAEC as well as before transmitting them to the far-end room. To see that this is the case, we can look at the normal equations the echo canceler has to solve,

$$\mathbf{R}_{\widetilde{\mathbf{x}}}\widetilde{\mathbf{h}} = \mathbf{p}_{\widetilde{\mathbf{x}}d}, \tag{1.1}$$

J. Benesty et al.: A Perspective on Stereophonic Acoustic Echo Cancellation, STSP 4, pp. 1–4.
springerlink.com

where $\mathbf{R}_{\widetilde{\mathbf{x}}}$ is the correlation matrix of the excitation (loudspeaker) signal (left and right stereo channels), $\widetilde{\mathbf{h}}$ is the estimated echo path, and $\mathbf{p}_{\widetilde{\mathbf{x}}d}$ is the cross-correlation between the excitation signal and the microphone signal. See Chapter 3 for more details of the problem formulation, notation, and normal equations. The estimated echo path is given by the solution to the normal equations (1.1), which is found to be

$$\widetilde{\mathbf{h}} = \widetilde{\mathbf{h}}_{\mathrm{t}} + \sum_{i=\mathcal{R}+1}^{2L} \beta_{\mathrm{f},i}\mathbf{q}_i, \tag{1.2}$$

where $\widetilde{\mathbf{h}}_{\mathrm{t}}$ is the true echo path of length $2L$, $\mathbf{q}_i$ are eigenvectors (corresponding to the nullspace of $\mathbf{R}_{\widetilde{\mathbf{x}}}$), $\mathcal{R}$ is the rank of the correlation matrix, and $\beta_{\mathrm{f},i}$ are arbitrary factors. This solution is easily shown to be valid by using (1.1) and (1.2),

$$\begin{aligned} \mathbf{R}_{\widetilde{\mathbf{x}}}\widetilde{\mathbf{h}} &= \mathbf{R}_{\widetilde{\mathbf{x}}}\widetilde{\mathbf{h}}_{\mathrm{t}} + \underbrace{\sum_{i=\mathcal{R}+1}^{2L} \beta_{\mathrm{f},i}\mathbf{R}_{\widetilde{\mathbf{x}}}\mathbf{q}_i}_{\mathbf{0}} \\ &= \mathbf{p}_{\widetilde{\mathbf{x}}d}. \end{aligned} \tag{1.3}$$

Note that the solution (1.2) is independent of any adaptive algorithm we use in our echo canceler system. Whatever adaptive algorithm used will end up with non-zero scalar-values ($\beta_{\mathrm{f},i}$). It is clearly seen that we can only achieve a unique solution if $\mathcal{R} = 2L$ and this condition can only be met if we modify (preprocess) the signals that actually excite the transmission room.

Having concluded that preprocessing of far-end loudspeaker signals that actually are transmitted to the near-end room is the only way to achieve a unique solution, we turn to the latter problem (b). There has been a common misunderstanding in several publications that manipulating the adaptive algorithm to improve convergence rate solve (a) which is not true. However, using an algorithm tailored to exploit the cross-correlation between the channels addresses problem (b), i.e., it mitigates the effects of the ill-conditioned normal equations to be solved. Remember, even with sophisticated preprocessing, it is difficult to achieve a well-conditioned system. Various algorithm choices for problem (b) have been presented in literature. For example, a natural choice of algorithm is the recursive least-squares (RLS) ([5], [6], [7]), which was the preferred algorithm in [1] and subsequent papers such as [8], [9].

In order to build a working echo cancellation system, it is crucial to control the adaptive algorithm properly during different talker conditions. Talker conditions usually include; single talk cases, i.e., only the far-end or near-end talker is active, double-talk where both talkers are active simultaneously, as well as the idle condition with neither side active. A number of control mechanisms are commonly employed to control the algorithms under these various conditions and one of the most important is the double-talk detector (DTD). The objective of the DTD is to stop algorithm divergence during double-talk. Its functionality can either be incorporated directly into the adaptive algorithm, e.g., as a step-size control mechanism, or as a separate control module. Because of its importance and the existing wealth of publications in this area a chapter in this book is solely devoted to this problem. Another equally important aspect of echo canceler systems is handling of residual echo, usually referred to residual echo suppression or nonlinear processing (NLP) (of the residual echo). In a realistic acoustic environment, linear cancellation can never provide sufficient echo attenuation in every talker condition. To handle loud echoes, e.g., at initial convergence, echo path changes, or large acoustic coupling conditions, echo suppression is required to complement the linear echo canceler. Aspects of combined residual echo and noise suppression is therefore presented as a separate chapter.

1.2 Organization of the Book

The objectives of this book are to recast the stereophonic echo cancellation problem using the widely linear (WL) model, as well as in this framework present and analyze some of the typical algorithms applied to the stereophonic case. Chapter 2 describes the stereophonic echo cancellation problem as a WL model and redefines some of the evaluation criteria commonly used in echo cancellation. General identification of the stereophonic echo paths using the Wiener formulation in the WL stereo framework is discussed in Chapter 3. This chapter also analyzes the nonuniqueness problem and presents a new approach to preprocessing the loudspeaker signals. Three chapters are devoted to classical as well as improved variants of adaptive filters for the SAEC problem. Stochastic gradient methods, of which the normalized least-mean-square (NLMS) algorithm belongs, is the topic of Chapter 4. This chapter also discusses in detail how to appropriately regularize the algorithms. Regularization is extremely important for practical implementations of echo cancelers. Moreover, variable step-size control for NLMS based algorithms are presented. For the stereophonic problem, the ability of the adaptive algorithm to exploit the spatial correlation between the channels is important. A family of algorithms with this ability is based on affine projections (APs). Chapter 5 goes into details of these algorithms. AP algorithms (APAs) have less degrees of freedom for spatial decorrelation compared to RLS based algorithms. However, the APA is less computationally complex compared to the RLS and is therefore an interesting alternative for realtime implementations. RLS adaptive filters are the most flexible algorithms when it comes to handling the problems occurring in stereophonic echo cancellation systems. Hence, a full derivation of the WL model-based RLS as well as a fast version are described in Chapter 6. The problems of double-talk and residual echo and noise handling are treated in Chapters 7 and 8, respectively. Chapter 9 presents extensive simulation results from most of the algorithms described in previous chapters.

References

1. M. M. Sondhi and D. R. Morgan, "Acoustic echo cancellation for stereophonic teleconferencing," in *Proc. IEEE WASPAA*, 1991.
2. M. M. Sondhi, D. R. Morgan, and J. L. Hall, "Stereophonic acoustic echo cancellation–An overview of the fundamental problem," *IEEE Signal Process. Lett.*, vol. 2, pp. 148–151, Aug. 1995.
3. F. Amand, A. Gilloire, and J. Benesty, "Identifying the true echo path impulse responses in stereophonic acoustic echo cancellation," in *Proc. EUSIPCO*, pp. 1119–1122, 1996.
4. J. Benesty, D. R. Morgan, and M. M. Sondhi, "A better understanding and an improved solution to the specific problems of stereophonic acoustic echo cancellation," *IEEE Trans. Speech, Audio Process.*, vol. 6, pp. 156–165, Mar. 1998.
5. J. Cioffi and T. Kailath, "Fast recursive-least-squares transversal filters for adaptive filtering," *IEEE Trans. Acoust., Speech, Signal Process.*, vol. 34, pp. 304–337, Apr. 1984.
6. M. G. Bellanger, *Adaptive Filters and Signal Analysis.* NY: Dekker, 1988.
7. M. G. Bellanger and P. A. Regalia, "The FRL-QR algorithm for adaptive filtering: the case of multichannel signal,"*Signal Process.*, vol. 22, pp. 115–126, Feb. 1991.
8. J. Benesty, F. Amand, A. Gilloire, and Y. Grenier, "Adaptive filtering algorithms for stereophonic acoustic echo cancellation," in *Proc. IEEE ICASSP*, pp. 3099–3102, 1995.
9. P. Eneroth, S. L. Gay, T. Gänsler, and J. Benesty, "A real-time stereophic acoustic subband echo canceler," in *Acoustic Signal Processing for Telecommunication*, S. L. Gay and J. Benesty, eds., Kluwer Academic Publishers, 2000, Chapter 8, pp. 135–153.

Chapter 2
Problem Formulation

It is well known that stereophonic acoustic echo, due to the coupling between two loudspeakers and two microphones, can be modelled by a two-input/two-output system with real random variables. In this chapter, we recast this problem as a single-input/single-output system with complex random variables by using the widely linear model. As a consequence, the four real-valued acoustic impulse responses are converted to one complex-valued impulse response. Also, all important conventional measures are reformulated in this new context. The main advantage of this approach is that instead of handling two (real) output signals separately, we only handle one (complex) output signal. This makes it convenient to handle the main three challenges of SAEC, i.e., system identification, double-talk detection, and echo suppression.

2.1 Stereophonic Acoustic Echo Model

In this work, we assume that all signals we deal with are zero mean. In the stereophonic setup, we have two input or loudspeaker signals denoted by $x_{\mathrm{L}}(n)$ and $x_{\mathrm{R}}(n)$ ("left" and "right"), and two output or microphone signals denoted by $d_{\mathrm{L}}(n)$ and $d_{\mathrm{R}}(n)$, which can be expressed as

$$d_{\mathrm{L}}(n) = y_{\mathrm{L}}(n) + v_{\mathrm{L}}(n), \tag{2.1}$$

$$d_{\mathrm{R}}(n) = y_{\mathrm{R}}(n) + v_{\mathrm{R}}(n), \tag{2.2}$$

where $y_{\mathrm{L}}(n)$ and $y_{\mathrm{R}}(n)$ are the stereo echo signals, and $v_{\mathrm{L}}(n)$ and $v_{\mathrm{R}}(n)$ are the near-end signals. Depending on the context, the near-end signals are either noise or a combination of noise and a near-end talker. Unless stated otherwise, we consider $v_{\mathrm{L}}(n)$ and $v_{\mathrm{R}}(n)$ additive noise signals. Obviously, $y_{\mathrm{L}}(n)$ and $y_{\mathrm{R}}(n)$ are independent of $v_{\mathrm{L}}(n)$ and $v_{\mathrm{R}}(n)$. The loudspeaker and microphone signals are all real random variables in the context of acoustic echo cancellation. Furthermore, we always model the echo signals as [1], [2]

J. Benesty et al.: A Perspective on Stereophonic Acoustic Echo Cancellation, STSP 4, pp. 5–11.
springerlink.com

$$y_{\mathrm{L}}(n) = \mathbf{h}_{\mathrm{t,LL}}^T \mathbf{x}_{\mathrm{L}}(n) + \mathbf{h}_{\mathrm{t,RL}}^T \mathbf{x}_{\mathrm{R}}(n), \quad (2.3)$$

$$y_{\mathrm{R}}(n) = \mathbf{h}_{\mathrm{t,LR}}^T \mathbf{x}_{\mathrm{L}}(n) + \mathbf{h}_{\mathrm{t,RR}}^T \mathbf{x}_{\mathrm{R}}(n), \quad (2.4)$$

where $\mathbf{h}_{\mathrm{t,LL}}, \mathbf{h}_{\mathrm{t,RL}}, \mathbf{h}_{\mathrm{t,LR}}, \mathbf{h}_{\mathrm{t,RR}}$ are L-dimensional vectors of the loudspeaker-to-microphone ("true") acoustic impulse responses, the superscript ^{T} denotes transpose of a vector or a matrix, and

$$\mathbf{x}_{\mathrm{L}}(n) = \left[x_{\mathrm{L}}(n) \; x_{\mathrm{L}}(n-1) \cdots x_{\mathrm{L}}(n-L+1) \right]^T$$

$$\mathbf{x}_{\mathrm{R}}(n) = \left[x_{\mathrm{R}}(n) \; x_{\mathrm{R}}(n-1) \cdots x_{\mathrm{R}}(n-L+1) \right]^T$$

are vectors comprising the L most recent loudspeaker signal samples.

We observe that the stereophonic acoustic echo is modelled by a two-input/two-output system. The aim is then to estimate the four acoustic impulse responses, $\mathbf{h}_{\mathrm{t,LL}}, \mathbf{h}_{\mathrm{t,RL}}, \mathbf{h}_{\mathrm{t,LR}}, \mathbf{h}_{\mathrm{t,RR}}$, from the microphone signals in order to cancel the echo due to the coupling between the loudspeakers and the microphones.

2.2 Widely Linear (WL) Model

Let us start by introducing the complex notation. From the two real random variables $d_{\mathrm{L}}(n)$ and $d_{\mathrm{R}}(n)$, we can form the complex random variable (CRV)

$$\begin{aligned} d(n) &= d_{\mathrm{L}}(n) + j d_{\mathrm{R}}(n) \\ &= y(n) + v(n), \end{aligned} \quad (2.5)$$

where $j = \sqrt{-1}$, $y(n) = y_{\mathrm{L}}(n) + j y_{\mathrm{R}}(n)$, and $v(n) = v_{\mathrm{L}}(n) + j v_{\mathrm{R}}(n)$.

Let us define the complex random vector

$$\mathbf{x}(n) = \mathbf{x}_{\mathrm{L}}(n) + j \mathbf{x}_{\mathrm{R}}(n). \quad (2.6)$$

We can express the (complex) echo signal as

$$y(n) = \mathbf{h}_{\mathrm{t}}^H \mathbf{x}(n) + \mathbf{h}_{\mathrm{t}}'^H \mathbf{x}^*(n), \quad (2.7)$$

where the superscripts ^{H} and * denote transpose-conjugate and conjugate, respectively, and

$$\mathbf{h}_{\mathrm{t}} = \mathbf{h}_{\mathrm{t},1} + j \mathbf{h}_{\mathrm{t},2}, \quad (2.8)$$

$$\mathbf{h}_{\mathrm{t}}' = \mathbf{h}_{\mathrm{t},1}' + j \mathbf{h}_{\mathrm{t},2}', \quad (2.9)$$

with

$$\mathbf{h}_{\mathrm{t},1} = \frac{\mathbf{h}_{\mathrm{t,LL}} + \mathbf{h}_{\mathrm{t,RR}}}{2}, \quad (2.10)$$

$$\mathbf{h}_{\mathrm{t},2} = \frac{\mathbf{h}_{\mathrm{t,RL}} - \mathbf{h}_{\mathrm{t,LR}}}{2}, \tag{2.11}$$

$$\mathbf{h}'_{\mathrm{t},1} = \frac{\mathbf{h}_{\mathrm{t,LL}} - \mathbf{h}_{\mathrm{t,RR}}}{2}, \tag{2.12}$$

$$\mathbf{h}'_{\mathrm{t},2} = -\frac{\mathbf{h}_{\mathrm{t,RL}} + \mathbf{h}_{\mathrm{t,LR}}}{2}. \tag{2.13}$$

We can rewrite (2.7) as

$$y(n) = \widetilde{\mathbf{h}}_{\mathrm{t}}^{H} \widetilde{\mathbf{x}}(n), \tag{2.14}$$

where

$$\widetilde{\mathbf{h}}_{\mathrm{t}} = \begin{bmatrix} \mathbf{h}_{\mathrm{t}} \\ \mathbf{h}'_{\mathrm{t}} \end{bmatrix},$$

$$\widetilde{\mathbf{x}}(n) = \begin{bmatrix} \mathbf{x}(n) \\ \mathbf{x}^*(n) \end{bmatrix}.$$

As a result, the complex observation is

$$d(n) = \widetilde{\mathbf{h}}_{\mathrm{t}}^{H} \widetilde{\mathbf{x}}(n) + v(n). \tag{2.15}$$

From (2.7) or (2.14), we recognize the widely linear (WL) model for CRVs proposed in [3], which is a nice generalization of the linear model for real random variables. Therefore, we see that we have now a complex acoustic impulse response, $\widetilde{\mathbf{h}}_{\mathrm{t}}$, whose complex input and output are, respectively, $x(n) = x_{\mathrm{L}}(n) + jx_{\mathrm{R}}(n)$ and $d(n)$.

Fundamentally, we have converted a two-input/two-output system with real random variables to a single-input/single-output system with CRVs thanks to the WL model. This approach is in line with the duality principle explained in [4]. The aim now is to estimate the complex acoustic impulse responses $\mathbf{h}_{\mathrm{t}}$ and $\mathbf{h}'_{\mathrm{t}}$ (or $\widetilde{\mathbf{h}}_{\mathrm{t}}$) from the complex observation, $d(n)$, and the complex input, $x(n)$. Figure 2.1 depicts the SAEC problem with the WL model, where $\widetilde{h}(n)$ is the adaptive filter.

Since we will mostly handle CRVs in the rest of this work, it is of interest to recall some useful definitions. A very important statistical characteristic of a CRV is the so-called circularity property or lack of it (noncircularity) [5], [6]. A zero-mean CRV, z, is circular if and only if the only nonnull moments and cumulants are the moments and cumulants constructed with the same power in z and z^* [7], [8]. In particular, z is said to be a second-order circular CRV (CCRV) if its so-called pseudo-variance [5] is equal to zero, i.e., $E\left(z^2\right) = 0$ with $E(\cdot)$ denoting mathematical expectation, while its variance is nonnull, i.e., $E\left(|z|^2\right) \neq 0$. This means that the second-order behavior of a CCRV is well described by its variance. If the pseudo-variance $E\left(z^2\right)$ is not equal to 0, then the CRV z is noncircular. A good measure of the second-order circularity is the circularity quotient [5] defined as the ratio between the pseudo-variance

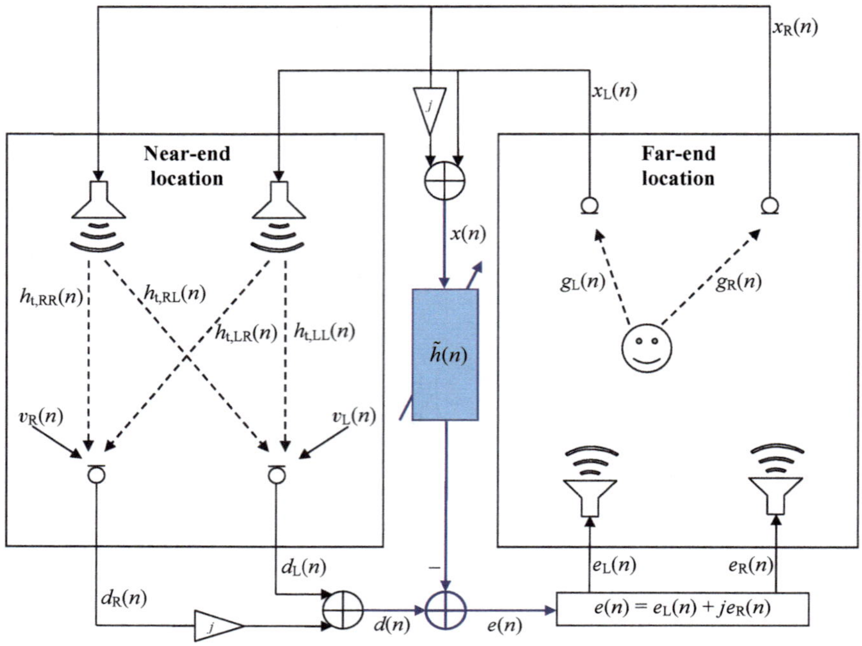

Fig. 2.1 Stereophonic acoustic echo cancellation (SAEC) with the widely linear (WL) model.

and the variance, i.e.,

$$\gamma_z = \frac{E\left(z^2\right)}{E\left(|z|^2\right)}. \tag{2.16}$$

It is easy to show that $0 \leq |\gamma_z| \leq 1$. If $\gamma_z = 0$, z is a second-order CCRV; otherwise, z is noncircular, and a larger value of $|\gamma_z|$ indicates that the CRV z is more noncircular.

Now, let us examine when a complex signal, $z = z_L + jz_R$, is second-order circular. We have

$$\gamma_z = \frac{E\left(z_L^2\right) - E\left(z_R^2\right) + 2jE\left(z_L z_R\right)}{\sigma_z^2}, \tag{2.17}$$

where $\sigma_z^2 = E\left(|z|^2\right)$ is the variance of z. One can check from (2.17) that the CRV z is second-order circular (i.e., $\gamma_z = 0$) if and only if

$$E\left(z_L^2\right) = E\left(z_R^2\right) \quad \text{and} \quad E\left(z_L z_R\right) = 0. \tag{2.18}$$

This means that the two real random variables z_{L} and z_{R} have the same variance and are uncorrelated.

2.3 Measures

In this section, we redefine in the context of the WL model some important measures used in echo cancellation.

We define the stereo echo-to-noise ratio (SENR)[1] as

$$\mathrm{SENR} = \frac{\sigma_y^2}{\sigma_v^2}, \tag{2.19}$$

where $\sigma_y^2 = E\left[|y(n)|^2\right]$ and $\sigma_v^2 = E\left[|v(n)|^2\right]$ are the variances of $y(n)$ and $v(n)$, respectively.

It is well known that acoustic impulse responses can be very sparse. One convenient way to measure this sparseness is via the sparseness measure given in [9], [10], [11], which can be extended to the complex case:

$$\mathcal{S}\left(\widetilde{\mathbf{h}}_{\mathrm{t}}\right) = \frac{2L}{2L - \sqrt{2L}}\left(1 - \frac{\left\|\widetilde{\mathbf{h}}_{\mathrm{t}}\right\|_1}{\sqrt{2L}\left\|\widetilde{\mathbf{h}}_{\mathrm{t}}\right\|_2}\right), \tag{2.20}$$

where

$$\|\mathbf{z}\|_1 = \sum_{l=1}^{2L} |z_l|$$

$$\|\mathbf{z}\|_2 = \sqrt{\sum_{l=1}^{2L} |z_l|^2}$$

are the ℓ_1 and ℓ_2 norms of the $2L$-dimensional complex vector

$$\mathbf{z} = \left[\, z_1 \; z_2 \; \cdots \; z_{2L} \,\right]^T.$$

It can be verified that $0 \leq \mathcal{S}\left(\widetilde{\mathbf{h}}_{\mathrm{t}}\right) \leq 1$ and $\mathcal{S}\left(a\widetilde{\mathbf{h}}_{\mathrm{t}}\right) = \mathcal{S}\left(\widetilde{\mathbf{h}}_{\mathrm{t}}\right)$, $\forall a \neq 0$. The larger the value of $\mathcal{S}\left(\widetilde{\mathbf{h}}_{\mathrm{t}}\right)$, the sparser the complex acoustic impulse response.

Let $\widetilde{\mathbf{h}}(n)$ be an adaptive filter, which is an estimate of $\widetilde{\mathbf{h}}_{\mathrm{t}}$, and let

$$\widehat{y}(n) = \widetilde{\mathbf{h}}^H(n-1)\widetilde{\mathbf{x}}(n) \tag{2.21}$$

[1] This definition is equivalent to the signal-to-noise ratio (SNR).

be the output of the adaptive filter at time n. An objective measure to assess the echo cancellation by the adaptive filter is the echo-return loss enhancement (ERLE) defined as [12]

$$\mathrm{ERLE}(n) = \frac{\sigma_y^2}{E\left[|y(n) - \widehat{y}(n)|^2\right]}. \tag{2.22}$$

Perhaps, the most used performance measure in echo cancellation is the so-called normalized misalignment [13]. It quantifies directly how "well" (in terms of convergence, tracking, and accuracy to the solution) an adaptive filter converges to the impulse response of the system that needs to be identified. The normalized misalignment in the WL context is defined as

$$\mathrm{Mis}(n) = \frac{\left\|\widetilde{\mathbf{h}}_{\mathrm{t}} - \widetilde{\mathbf{h}}(n)\right\|_2^2}{\left\|\widetilde{\mathbf{h}}_{\mathrm{t}}\right\|_2^2} \tag{2.23}$$

or in dB,

$$\mathrm{Mis}(n) = 20\log_{10}\frac{\left\|\widetilde{\mathbf{h}}_{\mathrm{t}} - \widetilde{\mathbf{h}}(n)\right\|_2}{\left\|\widetilde{\mathbf{h}}_{\mathrm{t}}\right\|_2} \quad \text{(dB)}. \tag{2.24}$$

References

1. M. M. Sondhi, D. R. Morgan, and J. L. Hall, "Stereophonic acoustic echo cancellation–An overview of the fundamental problem," *IEEE Signal Process. Lett.*, vol. 2, pp. 148–151, Aug. 1995.
2. J. Benesty, D. R. Morgan, and M. M. Sondhi, "A better understanding and an improved solution to the specific problems of stereophonic acoustic echo cancellation," *IEEE Trans. Speech, Audio Process.*, vol. 6, pp. 156–165, Mar. 1998.
3. B. Picinbono and P. Chevalier, "Widely linear estimation with complex data," *IEEE Trans. Signal Process.*, vol. 43, pp. 2030–2033, Aug. 1995.
4. D. P. Mandic, S. Still, and S. C. Douglas, "Duality between widely linear and dual channel adaptive filtering," in *Proc. IEEE ICASSP*, 2009, pp. 1729–1732.
5. E. Ollila, "On the circularity of a complex random variable," *IEEE Signal Process. Lett.*, vol. 15, pp. 841–844, 2008.
6. D. P. Mandic and S. L. Goh, *Complex Valued Nonlinear Adaptive Filters: Noncircularity, Widely Linear and Neural Models.* Wiley, 2009.
7. P. O. Amblard, M. Gaeta, and J. L. Lacoume, "Statistics for complex variables and signals–Part I: variables," *Signal Process.*, vol. 53, pp. 1–13, 1996.
8. P. O. Amblard, M. Gaeta, and J. L. Lacoume, "Statistics for complex variables and signals–Part II: signals," *Signal Process.*, vol. 53, pp. 15–25, 1996.
9. P. O. Hoyer, "Non-negative matrix factorization with sparseness constraints," *J. Machine Learning Res.*, vol. 49, pp. 1208–1215, June 2001.
10. Y. Huang, J. Benesty, and J. Chen, *Acoustic MIMO Signal Processing.* Berlin, Germany: Springer-Verlag, 2006.

11. C. Paleologu, J. Benesty, and S. Ciochină, *Sparse Adaptive Filters for Echo Cancellation.* San Rafael: Morgan & Claypool, 2010.
12. E. Hänsler and G. Schmidt, *Acoustic Echo and Noise Control–A Practical Approach.* Hoboken, NJ: Wiley, 2004.
13. J. Benesty, T. Gänsler, D. R. Morgan, M. M. Sondhi, and S. L. Gay, *Advances in Network and Acoustic Echo Cancellation.* Berlin, Germany: Springer-Verlag, 2001.

Chapter 3
System Identification with the Wiener Filter

The main objective of SAEC is to identify the four acoustic impulse responses, $\mathbf{h}_{t,LL}, \mathbf{h}_{t,RL}, \mathbf{h}_{t,LR}, \mathbf{h}_{t,RR}$, or, equivalently, the complex acoustic impulse response, $\widetilde{\mathbf{h}}_t$, of the stereophonic system. In this chapter, we show how to estimate $\widetilde{\mathbf{h}}_t$ with the Wiener approach [1], which has been extremely useful in many problems, in general, and in echo cancellation, in particular. The Wiener filter is derived from the mean-square error (MSE) criterion. We will discuss the well-known nonuniqueness problem that occurs in SAEC but reformulated in the WL model context. Because of this problem, some pre-processing of the complex input signal may be necessary. We also study, in the context of SAEC, the deterministic algorithm, which is an iterative approach to find the Wiener filter. Finally, we end this chapter by discussing the regularized MSE criterion, which can be very useful for the derivation of filters that promote sparsity. This approach has the potential to solve the SAEC problem without distorting the input signals.

3.1 Mean-Square Error (MSE) Criterion and Wiener Filter

With the Wiener theory, it is possible under some conditions to identify the impulse response $\widetilde{\mathbf{h}}_t$, given the input and output signals $x(n)$ and $d(n)$. Define the complex error signal

$$\begin{aligned} e(n) &= d(n) - \widehat{y}(n) \\ &= d(n) - \widetilde{\mathbf{h}}^H \widetilde{\mathbf{x}}(n), \end{aligned} \tag{3.1}$$

which is the difference between the output signal and the estimate of the (complex) echo, and where

J. Benesty et al.: A Perspective on Stereophonic Acoustic Echo Cancellation, STSP 4, pp. 13–27.
springerlink.com

$$\widetilde{\mathbf{h}} = \begin{bmatrix} \mathbf{h} \\ \mathbf{h}' \end{bmatrix}$$

is the estimate of $\widetilde{\mathbf{h}}_{\mathrm{t}}$ (with both vectors having the same length $2L$).

From (3.1), we can form the mean-square error (MSE) criterion defined as [2]

$$\begin{aligned} J\left(\widetilde{\mathbf{h}}\right) &= E\left[|e(n)|^2\right] \\ &= \sigma_d^2 - \widetilde{\mathbf{h}}^H \mathbf{p}_{\widetilde{\mathbf{x}}d} - \mathbf{p}_{\widetilde{\mathbf{x}}d}^H \widetilde{\mathbf{h}} + \widetilde{\mathbf{h}}^H \mathbf{R}_{\widetilde{\mathbf{x}}} \widetilde{\mathbf{h}} \\ &= \sigma_v^2 + \left(\widetilde{\mathbf{h}}_{\mathrm{t}} - \widetilde{\mathbf{h}}\right)^H \mathbf{R}_{\widetilde{\mathbf{x}}} \left(\widetilde{\mathbf{h}}_{\mathrm{t}} - \widetilde{\mathbf{h}}\right), \end{aligned} \tag{3.2}$$

where $\sigma_d^2 = E\left[|d(n)|^2\right]$ is the variance of $d(n)$,

$$\begin{aligned} \mathbf{p}_{\widetilde{\mathbf{x}}d} &= E\left[\widetilde{\mathbf{x}}(n) d^*(n)\right] \\ &= \mathbf{R}_{\widetilde{\mathbf{x}}} \widetilde{\mathbf{h}}_{\mathrm{t}} \end{aligned} \tag{3.3}$$

is the cross-correlation vector between $\widetilde{\mathbf{x}}(n)$ and $d(n)$, and

$$\mathbf{R}_{\widetilde{\mathbf{x}}} = E\left[\widetilde{\mathbf{x}}(n) \widetilde{\mathbf{x}}^H(n)\right] \tag{3.4}$$

is the covariance matrix of $\widetilde{\mathbf{x}}(n)$, which can be rewritten as

$$\mathbf{R}_{\widetilde{\mathbf{x}}} = \begin{bmatrix} \mathbf{R}_{\mathbf{x}} & \mathbf{R}_{\mathbf{x}\mathbf{x}^*} \\ \mathbf{R}_{\mathbf{x}\mathbf{x}^*}^H & \mathbf{R}_{\mathbf{x}}^T \end{bmatrix}, \tag{3.5}$$

where

$$\mathbf{R}_{\mathbf{x}} = E\left[\mathbf{x}(n) \mathbf{x}^H(n)\right] \tag{3.6}$$

and

$$\mathbf{R}_{\mathbf{x}\mathbf{x}^*} = E\left[\mathbf{x}(n) \mathbf{x}^T(n)\right] \tag{3.7}$$

are, respectively, the covariance and pseudo-covariance matrices of $\mathbf{x}(n)$. The pseudo-covariance matrix of $\mathbf{x}(n)$ is also the cross-correlation matrix between $\mathbf{x}(n)$ and $\mathbf{x}^*(n)$. In the particular case where $\mathbf{x}_{\mathrm{L}}(n)$ and $\mathbf{x}_{\mathrm{R}}(n)$ are uncorrelated, the covariance matrix $\mathbf{R}_{\widetilde{\mathbf{x}}}$ reduces to a real-valued matrix

$$\mathbf{R}_{\widetilde{\mathbf{x}}} = \begin{bmatrix} \mathbf{R}_{\mathbf{x}_{\mathrm{L}}} + \mathbf{R}_{\mathbf{x}_{\mathrm{R}}} & \mathbf{R}_{\mathbf{x}_{\mathrm{L}}} - \mathbf{R}_{\mathbf{x}_{\mathrm{R}}} \\ \mathbf{R}_{\mathbf{x}_{\mathrm{L}}} - \mathbf{R}_{\mathbf{x}_{\mathrm{R}}} & \mathbf{R}_{\mathbf{x}_{\mathrm{L}}} + \mathbf{R}_{\mathbf{x}_{\mathrm{R}}} \end{bmatrix}, \tag{3.8}$$

where $\mathbf{R}_{\mathbf{x}_{\mathrm{L}}} = E\left[\mathbf{x}_{\mathrm{L}}(n) \mathbf{x}_{\mathrm{L}}^T(n)\right]$ and $\mathbf{R}_{\mathbf{x}_{\mathrm{R}}} = E\left[\mathbf{x}_{\mathrm{R}}(n) \mathbf{x}_{\mathrm{R}}^T(n)\right]$.

The optimal Wiener filter, $\widetilde{\mathbf{h}}_{\mathrm{W}}$, is the one that cancels the gradient of $J\left(\widetilde{\mathbf{h}}\right)$ with respect to $\widetilde{\mathbf{h}}^H$, i.e.,

$$\frac{\partial J\left(\widetilde{\mathbf{h}}\right)}{\partial \widetilde{\mathbf{h}}^H} = \mathbf{0}. \tag{3.9}$$

We have

$$\begin{aligned}\frac{\partial J\left(\widetilde{\mathbf{h}}\right)}{\partial \widetilde{\mathbf{h}}^H} &= E\left[e^*(n)\frac{\partial e(n)}{\partial \widetilde{\mathbf{h}}^H}\right] \\ &= -E\left[e^*(n)\widetilde{\mathbf{x}}(n)\right].\end{aligned} \tag{3.10}$$

Therefore, at the optimum, we have

$$E\left[e_{\mathrm{W}}^*(n)\widetilde{\mathbf{x}}(n)\right] = \mathbf{0}, \tag{3.11}$$

where

$$e_{\mathrm{W}}(n) = d(n) - \widetilde{\mathbf{h}}_{\mathrm{W}}^H\widetilde{\mathbf{x}}(n) \tag{3.12}$$

is the error signal for which $J\left(\widetilde{\mathbf{h}}\right)$ is minimized (i.e., the optimal filter). Expression (3.11) is called the orthogonality principle.

The optimal estimate of the (complex) echo, $y(n)$, is then

$$\widehat{y}_{\mathrm{W}}(n) = \widetilde{\mathbf{h}}_{\mathrm{W}}^H\widetilde{\mathbf{x}}(n). \tag{3.13}$$

It is easy to check, with the help of the orthogonality principle, that we also have

$$E\left[e_{\mathrm{W}}^*(n)\widehat{y}_{\mathrm{W}}(n)\right] = 0. \tag{3.14}$$

The previous expression is called the corollary to the orthogonality principle.

If we substitute (3.12) into (3.11), we find the linear system of $2L$ equations, which are also known as the Wiener-Hopf (or normal) equations:

$$\mathbf{R}_{\widetilde{\mathbf{x}}}\widetilde{\mathbf{h}}_{\mathrm{W}} = \mathbf{p}_{\widetilde{\mathbf{x}}d}. \tag{3.15}$$

Assuming that $\mathbf{R}_{\widetilde{\mathbf{x}}}$ is non-singular, the optimal Wiener filter is then

$$\begin{aligned}\widetilde{\mathbf{h}}_{\mathrm{W}} &= \mathbf{R}_{\widetilde{\mathbf{x}}}^{-1}\mathbf{p}_{\widetilde{\mathbf{x}}d} \\ &= \widetilde{\mathbf{h}}_{\mathrm{t}}.\end{aligned} \tag{3.16}$$

Solving (3.16) gives exactly the complex impulse response of the system.

The criterion $J\left(\widetilde{\mathbf{h}}\right)$ is a quadratic function of the filter coefficient vector $\widetilde{\mathbf{h}}$ and has a single minimum point (assuming that $\mathbf{R}_{\widetilde{\mathbf{x}}}$ is non-singular). This point combines the optimal Wiener filter, as shown above, and a value called the minimum MSE (MMSE), which is obtained by substituting (3.16) into

(3.2):

$$\begin{aligned} J_{\min} &= J\left(\widetilde{\mathbf{h}}_{\mathrm{W}}\right) \\ &= \sigma_d^2 - \widetilde{\mathbf{h}}_{\mathrm{W}}^H \mathbf{R}_{\widetilde{\mathbf{x}}} \widetilde{\mathbf{h}}_{\mathrm{W}} \\ &= \sigma_d^2 - \sigma_{\widehat{y}_{\mathrm{W}}}^2, \end{aligned} \tag{3.17}$$

where

$$\sigma_{\widehat{y}_{\mathrm{W}}}^2 = E\left[\left|\widehat{y}_{\mathrm{W}}(n)\right|^2\right] \tag{3.18}$$

is the variance of the optimal filter output signal, $\widehat{y}_{\mathrm{W}}(n)$. This MMSE can be rewritten as

$$J_{\min} = \sigma_v^2. \tag{3.19}$$

We define the normalized MMSE (NMMSE) as

$$\begin{aligned} J_{\mathrm{N,min}} &= \frac{J_{\min}}{\sigma_d^2} \\ &= \frac{1}{1 + \mathrm{SENR}} \leq 1. \end{aligned} \tag{3.20}$$

The previous expression shows how the NMMSE is related to the SENR.

3.2 Nonuniqueness Problem

In the single-channel acoustic echo cancellation problem, the covariance matrix that needs to be inverted in the Wiener-Hopf equations is usually full-rank, although it can be ill-conditioned. However, it is well known that in the SAEC problem, most of the time, the two input signals $x_{\mathrm{L}}(n)$ and $x_{\mathrm{R}}(n)$ are obtained by filtering a common source, so that a problem of nonuniqueness is expected [3]. In this scenario, we have the following relation [4], [5]

$$\mathbf{x}_{\mathrm{L}}^T(n)\mathbf{g}_{\mathrm{R}} = \mathbf{x}_{\mathrm{R}}^T(n)\mathbf{g}_{\mathrm{L}}, \tag{3.21}$$

where $\mathbf{g}_{\mathrm{L}}$ and $\mathbf{g}_{\mathrm{R}}$ are the L-dimensional vectors of the source-to-microphone acoustic impulse responses in the far-end room.

Define the two complex vectors

$$\begin{aligned} \mathbf{g} &= \mathbf{g}_{\mathrm{L}} + j\mathbf{g}_{\mathrm{R}}, \\ \widetilde{\mathbf{g}} &= \begin{bmatrix} \mathbf{g} \\ -\mathbf{g}^* \end{bmatrix}. \end{aligned}$$

It can be verified that

$$\mathbf{x}^H(n)\mathbf{g} = \mathbf{x}^T(n)\mathbf{g}^*. \tag{3.22}$$

As a result, we have

$$\mathbf{R}_{\widetilde{\mathbf{x}}}\widetilde{\mathbf{g}} = \mathbf{0}, \tag{3.23}$$

which means that the matrix $\mathbf{R}_{\widetilde{\mathbf{x}}}$ is not invertible. In fact, $\widetilde{\mathbf{g}}$ represents the nullspace of $\mathbf{R}_{\widetilde{\mathbf{x}}}$. Since we have only one linear relation, the dimension of this nullspace is equal to 1. Therefore, the rank of $\mathbf{R}_{\widetilde{\mathbf{x}}}$ is equal to $2L - 1$. Thus, there is no unique solution to the problem and an iterative/adaptive algorithm will drive to any one of many possible solutions, which can be very different from the "true" desired solution $\widetilde{\mathbf{h}} = \widetilde{\mathbf{h}}_{\mathrm{t}}$. These nonunique solutions are dependent on the impulse responses in the far-end room, i.e.,

$$\widetilde{\mathbf{h}} = \widetilde{\mathbf{h}}_{\mathrm{t}} + \beta_{\mathrm{f}}\widetilde{\mathbf{g}}, \tag{3.24}$$

where β_{f} is an arbitrary factor. This, of course, is intolerable because $\widetilde{\mathbf{g}}$ can change instantaneously, for example, as one person stops talking and another starts. In this case, the iterative/adaptive algorithm would have to track the changes in the far-field system, which can be overwhelming for the algorithm as it already has to track the changes in the near-end system.

Substituting (3.24) into (3.2) leads to

$$J\left(\widetilde{\mathbf{h}}\right) = J_{\min} = \sigma_v^2, \tag{3.25}$$

which means that all these nonunique solutions can cancel the stereo echo but the system may not be very stable.

3.3 Distortion for a Unique Solution

In order to have a unique solution to the SAEC problem, it may be required to distort the input signals $x_{\mathrm{L}}(n)$ and $x_{\mathrm{R}}(n)$. A distortion that reduces the coherence between these two signals will lead to the estimation of the true acoustic impulse responses [7]. However, this distortion must be performed in such a way that the quality of the signals and the stereo effect are not degraded. It is well known that the magnitude coherence between two processes is equal to 1 if and only if they are linearly related. In order to weaken this relation, some nonlinear or time-varying transformation of the stereo channels has to be made.

A simple nonlinear method that gives good performances uses a (positive) half-wave rectifier [5]. The nonlinearly transformed signals become

$$x'_{\mathrm{L}}(n) = x_{\mathrm{L}}(n) + \alpha_{\mathrm{r}} \frac{x_{\mathrm{L}}(n) + |x_{\mathrm{L}}(n)|}{2}, \tag{3.26}$$

$$x'_{\mathrm{R}}(n) = x_{\mathrm{R}}(n) + \alpha_{\mathrm{r}} \frac{x_{\mathrm{R}}(n) + |x_{\mathrm{R}}(n)|}{2}, \tag{3.27}$$

where α_{r} is a parameter used to control the amount of nonlinearity. For this method, there can only be a linear relation between the nonlinearly transformed channels if $\forall n$, $x_{\mathrm{L}}(n) \geq 0$ and $x_{\mathrm{R}}(n) \geq 0$ or if we have $ax_{\mathrm{L}}(n - n_{\mathrm{L}}) = x_{\mathrm{R}}(n - n_{\mathrm{R}})$ with $a > 0$. In practice, however, these cases almost never occur because we always have zero-mean signals and $\mathbf{g}_{\mathrm{L}}, \mathbf{g}_{\mathrm{R}}$ are rarely related by just a simple delay.

An improved version of this technique is to use positive and negative half-wave rectifiers on each channel respectively [5], i.e.,

$$x'_{\mathrm{L}}(n) = x_{\mathrm{L}}(n) + \alpha_{\mathrm{r}} \frac{x_{\mathrm{L}}(n) + |x_{\mathrm{L}}(n)|}{2}, \tag{3.28}$$

$$x'_{\mathrm{R}}(n) = x_{\mathrm{R}}(n) + \alpha_{\mathrm{r}} \frac{x_{\mathrm{R}}(n) - |x_{\mathrm{R}}(n)|}{2}. \tag{3.29}$$

This principle removes the linear relation even in the special signal cases given above.

Experiments show that stereo perception is not affected by the above methods even with α_{r} as large as 0.5. Also, the audible distortion introduced for speech is small because of the nature of the speech signal and psychoacoustic masking effects [8].

Other types of nonlinearities have also been investigated and compared [9]. The results indicate that, of the several nonlinearities considered, ideal half-wave rectification and smoothed half-wave rectification appear to be the best choices for speech. For music, the nonlinearity parameter of the ideal rectifier must be readjusted. The smoothed rectifier does not require this readjustment but is a little more complicated to implement.

We now propose a new distortion that fits well with the WL model.

We can express the complex input signal as

$$\begin{aligned} x(n) &= x_{\mathrm{L}}(n) + jx_{\mathrm{R}}(n) \\ &= e^{j\theta_{\mathrm{r}}(n)} |x(n)|, \end{aligned} \tag{3.30}$$

where $\theta_{\mathrm{r}}(n)$ [with $\tan\theta_{\mathrm{r}}(n) = x_{\mathrm{R}}(n)/x_{\mathrm{L}}(n)$] and $|x(n)| = \sqrt{x_{\mathrm{L}}^2(n) + x_{\mathrm{R}}^2(n)}$ are the phase and module of $x(n)$, respectively. In this formulation, we represent the stereo perception with $\theta_{\mathrm{r}}(n)$ and the quality of the stereo signals with $|x(n)|$. A modification of $\theta_{\mathrm{r}}(n)$ only, will mostly affect the stereo effect of $x(n)$; while a modification of $|x(n)|$ will mostly affect the quality of the stereo signals.

With the complex notation, (3.28)–(3.29) can be expressed as

$$x'(n) = x'_{\mathrm{L}}(n) + jx'_{\mathrm{R}}(n)$$

$$= e^{j\theta'_{\mathrm{r}}(n)} \left| x'(n) \right|, \tag{3.31}$$

where

$$\begin{aligned} \tan\theta'_{\mathrm{r}}(n) &= \frac{x'_{\mathrm{R}}(n)}{x'_{\mathrm{L}}(n)} \\ &= \tan\theta_{\mathrm{r}}(n) \cdot \frac{\alpha_{\mathrm{r}} + 2 + \alpha_{\mathrm{r}} \cdot \mathrm{sgn}\left[x_{\mathrm{L}}(n)\right]}{\alpha_{\mathrm{r}} + 2 - \alpha_{\mathrm{r}} \cdot \mathrm{sgn}\left[x_{\mathrm{R}}(n)\right]} \end{aligned} \tag{3.32}$$

and

$$\begin{aligned} &\left| x'(n) \right| = \\ &\sqrt{\left(1 + \alpha_{\mathrm{r}} + 0.5\alpha_{\mathrm{r}}^2\right) \left| x(n) \right|^2 + \alpha_{\mathrm{r}} \left(1 + 0.5\alpha_{\mathrm{r}}^2\right) \left[x_{\mathrm{L}}(n) \left| x_{\mathrm{L}}(n) \right| - x_{\mathrm{R}}(n) \left| x_{\mathrm{R}}(n) \right|\right]}. \end{aligned} \tag{3.33}$$

From the two previous expressions, we observe that both the phase and module are modified with a nonlinear distortion. Amazingly, even with a value of α_{r} as large as 0.5, the stereo effect is not affected. This is likely due to the fact that the phase is not changed randomly, like in some other approaches, but according to the changes of the stereo signals.

The SAEC problem happens because the signals $x_{\mathrm{L}}(n)$ and $x_{\mathrm{R}}(n)$ are linearly related. Let us consider the worst case scenario, where $x_{\mathrm{L}}(n)$ is equal to $x_{\mathrm{R}}(n)$, i.e.,

$$x_{\mathrm{L}}(n) = x_{\mathrm{R}}(n), \ \forall n. \tag{3.34}$$

In this situation, (3.31) becomes

$$x'(n) = \sqrt{1 + \alpha_{\mathrm{r}} + 0.5\alpha_{\mathrm{r}}^2} \cdot e^{j\theta'_{\mathrm{r}}(n)} \left| x(n) \right|, \tag{3.35}$$

where

$$\tan\theta'_{\mathrm{r}}(n) = (\alpha_{\mathrm{r}} + 1) \tan\theta_{\mathrm{r}}(n) \text{ if } x_{\mathrm{L}}(n) > 0 \tag{3.36}$$

and

$$\tan\theta'_{\mathrm{r}}(n) = \frac{1}{\alpha_{\mathrm{r}} + 1} \tan\theta_{\mathrm{r}}(n) \text{ if } x_{\mathrm{L}}(n) < 0. \tag{3.37}$$

We see that the module is not affected since α_{r} is constant across time but $\theta'_{\mathrm{r}}(n)$ depends on $x_{\mathrm{L}}(n) = x_{\mathrm{R}}(n)$. As a result, only the phase is changed. While $x_{\mathrm{L}}(n) = x_{\mathrm{R}}(n)$, $x'_{\mathrm{L}}(n) \neq x'_{\mathrm{R}}(n)$ and the transformed signals are no more linearly related. We know by experience that, even in this difficult scenario, the misalignment is improved with the nonlinear transformations. This suggests that we may not really need to modify the module of the complex signal, $x(n)$. Therefore, we propose the new following transformations:

$$x''_{\mathrm{L}}(n) = \cos\theta'_{\mathrm{r}}(n) \left| x(n) \right|, \tag{3.38}$$

$$x''_{\mathrm{R}}(n) = \sin\theta'_{\mathrm{r}}(n)\,|x(n)|\,. \tag{3.39}$$

Clearly, the phase is computed from the half-wave rectifiers [eq. (3.32)] while the module corresponds to the module of the original signals. As a consequence, with (3.38)–(3.39) we may have the same misalignment as with (3.28)–(3.29) but with the advantage of little distortion. So we can even increase the value of α_{r} to have better performance as long as the stereo effect is not much affected. There are several other ideas that can be developed from this one.

3.4 Deterministic Algorithm

The deterministic or steepest-descent algorithm is actually an iterative algorithm of great importance since it is the starting point of adaptive filters. It is summarized by the simple recursion

$$\begin{aligned}\widetilde{\mathbf{h}}(n) &= \widetilde{\mathbf{h}}(n-1) - \mu \cdot \frac{\partial J\left[\widetilde{\mathbf{h}}(n-1)\right]}{\partial \widetilde{\mathbf{h}}^H(n-1)} \\ &= \widetilde{\mathbf{h}}(n-1) + \mu\left[\mathbf{p}_{\widetilde{\mathbf{x}}d} - \mathbf{R}_{\widetilde{\mathbf{x}}}\widetilde{\mathbf{h}}(n-1)\right],\ n \geq 1,\ \widetilde{\mathbf{h}}(0) = \mathbf{0},\end{aligned} \tag{3.40}$$

where μ is a positive constant called the step-size parameter. In this algorithm, $\mathbf{p}_{\widetilde{\mathbf{x}}d}$ and $\mathbf{R}_{\widetilde{\mathbf{x}}}$ are supposed to be known; and clearly the inversion of the matrix $\mathbf{R}_{\widetilde{\mathbf{x}}}$, which can be costly, is not needed. The deterministic algorithm can be reformulated with the error signal as

$$e(n) = d(n) - \widetilde{\mathbf{h}}^H(n-1)\widetilde{\mathbf{x}}(n), \tag{3.41}$$

$$\widetilde{\mathbf{h}}(n) = \widetilde{\mathbf{h}}(n-1) + \mu E[\widetilde{\mathbf{x}}(n)e^*(n)]. \tag{3.42}$$

Now the important question is: what are the conditions on μ to make the algorithm converge to the true complex acoustic impulse response $\widetilde{\mathbf{h}}_{\mathrm{t}}$? To answer this question, we will examine the natural modes of the algorithm [6].

We define the misalignment vector as

$$\mathbf{m}(n) = \widetilde{\mathbf{h}}_{\mathrm{t}} - \widetilde{\mathbf{h}}(n), \tag{3.43}$$

which is the difference between the complex impulse response of the system and the estimated one at iteration n. Injecting (3.3) in (3.40) and subtracting $\widetilde{\mathbf{h}}_{\mathrm{t}}$ on both sides of the equation, we obtain

$$\mathbf{m}(n) = \left(\mathbf{I}_{2L} - \mu\mathbf{R}_{\widetilde{\mathbf{x}}}\right)\mathbf{m}(n-1), \tag{3.44}$$

where $\mathbf{I}_{2L}$ is the $2L \times 2L$ identity matrix. Using the eigendecomposition of

$$\mathbf{R}_{\widetilde{\mathbf{x}}} = \mathbf{Q}\boldsymbol{\Lambda}\mathbf{Q}^H \tag{3.45}$$

in (3.44), where

$$\mathbf{Q}^H\mathbf{Q} = \mathbf{Q}\mathbf{Q}^H = \mathbf{I}_{2L}, \tag{3.46}$$

$$\boldsymbol{\Lambda} = \text{diag}\left(\lambda_0, \lambda_1, \ldots, \lambda_{2L-1}\right), \tag{3.47}$$

and $0 < \lambda_0 \leq \lambda_1 \leq \cdots \leq \lambda_{2L-1}$[1], we get the equivalent form

$$\mathbf{t}(n) = (\mathbf{I}_{2L} - \mu\boldsymbol{\Lambda})\mathbf{t}(n-1), \tag{3.48}$$

where

$$\begin{aligned} \mathbf{t}(n) &= \mathbf{Q}^H\mathbf{m}(n) \\ &= \mathbf{Q}^H\left[\widetilde{\mathbf{h}}_\mathrm{t} - \widetilde{\mathbf{h}}(n)\right]. \end{aligned} \tag{3.49}$$

Thus, for the lth natural mode of the steepest-descent algorithm, we have [2]

$$t_l(n) = (1 - \mu\lambda_l)t_l(n-1), \ l = 0, 1, \ldots, 2L-1, \tag{3.50}$$

or, equivalently,

$$t_l(n) = (1 - \mu\lambda_l)^n \, t_l(0), \ l = 0, 1, \ldots, 2L-1. \tag{3.51}$$

The algorithm converges to the true impulse response if

$$\lim_{n\to\infty} t_l(n) = 0, \ \forall l. \tag{3.52}$$

In this case,

$$\lim_{n\to\infty} \widetilde{\mathbf{h}}(n) = \widetilde{\mathbf{h}}_\mathrm{t}. \tag{3.53}$$

It is straightforward to see from (3.51) that a necessary and sufficient condition for the stability of the deterministic algorithm is that

$$-1 < 1 - \mu\lambda_l < 1, \ \forall l, \tag{3.54}$$

which implies

$$0 < \mu < \frac{2}{\lambda_l}, \ \forall l, \tag{3.55}$$

or

$$0 < \mu < \frac{2}{\lambda_{\max}}, \tag{3.56}$$

---

[1] We will not always assume that $\mathbf{R}_{\widetilde{\mathbf{x}}} > 0$.

where $\lambda_{\max}$ is the largest eigenvalue of the covariance matrix $\mathbf{R}_{\widetilde{\mathbf{x}}}$.

Now, let us see what happens when the signals $x_{\mathrm{L}}(n)$ and $x_{\mathrm{R}}(n)$ are linearly related. In this case, $\lambda_0 = 0$ and $\lambda_l > 0$, $\forall l \neq 0$. Using the stability condition given in (3.56) and taking $n \to \infty$, we find from (3.51) that

$$t_0(\infty) = t_0(0) = \mathbf{q}_0^H \widetilde{\mathbf{h}}_{\mathrm{t}}, \tag{3.57}$$

$$t_l(\infty) = 0, \ \forall l \neq 0, \tag{3.58}$$

where $\mathbf{q}_0$ is the eigenvector corresponding to the eigenvalue $\lambda_0 = 0$. As a result,

$$\begin{aligned}\widetilde{\mathbf{h}}(\infty) &= \widetilde{\mathbf{h}}_{\mathrm{t}} - \mathbf{Q}\mathbf{t}(\infty) \\ &= \widetilde{\mathbf{h}}_{\mathrm{t}} - \left(\mathbf{q}_0^H \widetilde{\mathbf{h}}_{\mathrm{t}}\right) \mathbf{q}_0 \\ &= \widetilde{\mathbf{h}}_{\mathrm{t}} + \beta_{\mathrm{f}} \widetilde{\mathbf{g}}.\end{aligned} \tag{3.59}$$

As expected, we obtain one of the nonunique solutions discussed in Section 3.2.

Assume that $\lambda_l > 0$, $\forall l$ and let us evaluate the time needed for each natural mode to converge to a given value. Expression (3.51) gives

$$\ln \frac{t_l(n)|}{t_l(0)|} = n \ln |1 - \mu\lambda_l|, \tag{3.60}$$

hence,

$$n = \frac{1}{\ln |1 - \mu\lambda_l|} \ln \frac{|t_l(n)|}{|t_l(0)|}. \tag{3.61}$$

The time constant, τ_l, for the lth natural mode is defined by taking $|t_l(n)|/|t_l(0)| = 1/e$ (where e is the base of the natural logarithm) in (3.61). Therefore,

$$\tau_l = \frac{-1}{\ln |1 - \mu\lambda_l|}. \tag{3.62}$$

We can link the time constant with the condition number of the correlation matrix $\mathbf{R}_{\widetilde{\mathbf{x}}}$. First, let

$$\mu = \frac{\alpha}{\lambda_{\max}}, \tag{3.63}$$

where

$$0 < \alpha < 2 \tag{3.64}$$

to guaranty the convergence of the algorithm and α is called the normalized step-size parameter. The smallest eigenvalue is $\lambda_{\min} = \lambda_0$; in this case,

$$\begin{aligned}\tau_0 &= \frac{-1}{\ln|1-\alpha\lambda_{\min}/\lambda_{\max}|} \\ &= \frac{-1}{\ln\left|1-\alpha/\chi\left(\mathbf{R}_{\widetilde{\mathbf{x}}}\right)\right|},\end{aligned} \tag{3.65}$$

where $\chi\left(\mathbf{R}_{\widetilde{\mathbf{x}}}\right) = \lambda_{\max}/\lambda_{\min}$ is the condition number of the matrix $\mathbf{R}_{\widetilde{\mathbf{x}}}$. We see that the convergence time of the slowest natural mode depends on the conditioning of $\mathbf{R}_{\widetilde{\mathbf{x}}}$.

From (3.49), we deduce the transient behavior of the misalignment:

$$\begin{aligned}\mathcal{H}(n) &= \mathbf{m}^H(n)\mathbf{m}(n) \\ &= \mathbf{t}^H(n)\mathbf{t}(n) \\ &= \left\|\widetilde{\mathbf{h}}_{\mathrm{t}} - \widetilde{\mathbf{h}}(n)\right\|_2^2 \\ &= \sum_{l=0}^{2L-1}(1-\mu\lambda_l)^{2n}|t_l(0)|^2.\end{aligned} \tag{3.66}$$

This value gives an idea on the global convergence of the filter to the true impulse response. This convergence is clearly governed by the smallest eigenvalues of $\mathbf{R}_{\widetilde{\mathbf{x}}}$. When $\mathbf{R}_{\widetilde{\mathbf{x}}} > 0$, $\mathcal{H}(\infty) = 0$ but when $\lambda_0 = 0$, $\mathcal{H}(\infty) = \left\|\mathbf{q}_0^H\widetilde{\mathbf{h}}_{\mathrm{t}}\right\|_2^2$ and the misalignment can never go to zero (i.e., the impulse response cannot be identified). A plot of $\mathcal{H}(n)$ versus n is called the learning curve of the misalignment.

We now examine the transient behavior of the MSE. Using $d(n) = \widetilde{\mathbf{h}}_{\mathrm{t}}^H\widetilde{\mathbf{x}}(n) + v(n)$, the error signal (3.41) can be rewritten as

$$\begin{aligned}e(n) &= d(n) - \widetilde{\mathbf{h}}^H(n-1)\widetilde{\mathbf{x}}(n) \\ &= v(n) + \mathbf{m}^H(n-1)\widetilde{\mathbf{x}}(n),\end{aligned} \tag{3.67}$$

so that the MSE is

$$\begin{aligned}J(n) &= E\left[|e(n)|^2\right] \\ &= \sigma_v^2 + \mathbf{m}^H(n-1)\mathbf{R}_{\widetilde{\mathbf{x}}}\mathbf{m}(n-1) \\ &= \sigma_v^2 + \mathbf{t}^H(n-1)\mathbf{\Lambda}\mathbf{t}(n-1) \\ &= \sigma_v^2 + \sum_{l=0}^{2L-1}\lambda_l(1-\mu\lambda_l)^{2n}|t_l(0)|^2.\end{aligned} \tag{3.68}$$

A plot of $J(n)$ versus n is called the learning curve of the MSE. Note that the MSE decays exponentially. When the algorithm is convergent, we see that

$$\lim_{n\to\infty} J(n) = \sigma_v^2. \tag{3.69}$$

This value corresponds to the MMSE, $J_{\min}$, obtained with the optimal Wiener filter. From (3.68), we observe that even when $\lambda_0 = 0$, the MSE converges to the variance of the noise, showing again that the echo can be cancelled while the impulse response cannot be identified.

Finally, to end this section, it is worth mentioning that a generalization of the deterministic algorithm is the Newton algorithm

$$\begin{aligned}\widetilde{\mathbf{h}}(n) &= \widetilde{\mathbf{h}}(n-1) - \left\{\frac{\partial^2 J\left[\widetilde{\mathbf{h}}(n-1)\right]}{\partial \widetilde{\mathbf{h}}^H(n-1)\partial \widetilde{\mathbf{h}}(n-1)}\right\}^{-1} \frac{\partial J\left[\widetilde{\mathbf{h}}(n-1)\right]}{\partial \widetilde{\mathbf{h}}^H(n-1)} \\ &= \mathbf{R}_{\widetilde{\mathbf{x}}}^{-1}\mathbf{p}_{\widetilde{\mathbf{x}}d},\end{aligned} \tag{3.70}$$

which converges in one iteration to the optimal Wiener filter (assuming that $\mathbf{R}_{\widetilde{\mathbf{x}}}$ is full rank). Needless to say that the Newton algorithm is never used in echo cancellation.

3.5 Regularized MSE Criterion

A more general criterion is the following:

$$\begin{aligned}J_{\text{reg},p}\left(\widetilde{\mathbf{h}}\right) &= J\left(\widetilde{\mathbf{h}}\right) + \delta\left\|\widetilde{\mathbf{h}}\right\|_p^p \\ &= E\left[|e(n)|^2\right] + \delta\left\|\widetilde{\mathbf{h}}\right\|_p^p,\end{aligned} \tag{3.71}$$

where $\delta \geq 0$ is the regularization parameter and $\|\cdot\|_p$ denotes the ℓ_p norm of a vector.

If the covariance matrix $\mathbf{R}_{\widetilde{\mathbf{x}}}$ is full rank but ill-conditioned, it is appropriate to choose the ℓ_2 norm. In this case, it is easy to see that the optimal filter is

$$\widetilde{\mathbf{h}} = \left(\mathbf{R}_{\widetilde{\mathbf{x}}} + \delta\mathbf{I}_{2L}\right)^{-1}\mathbf{p}_{\widetilde{\mathbf{x}}d}. \tag{3.72}$$

In practice, in the presence of noise and when $\mathbf{R}_{\widetilde{\mathbf{x}}}$ and $\mathbf{p}_{\widetilde{\mathbf{x}}d}$ have to be estimated, the filter given in (3.72) is much more reliable and accurate (with an adequate choice of δ) than the Wiener filter [eq. (3.16)] since $\mathbf{R}_{\widetilde{\mathbf{x}}} + \delta\mathbf{I}_{2L}$ is much better conditioned than $\mathbf{R}_{\widetilde{\mathbf{x}}}$.

In the SAEC problem, the matrix $\mathbf{R}_{\widetilde{\mathbf{x}}}$ can be rank deficient. So the classical regularization with the ℓ_2 norm will not lead to the solution we are looking for (i.e., $\widetilde{\mathbf{h}}_t$) but to one of the many possible solutions discussed in Section 3.2. Indeed, (3.72) can be rewritten as

$$\begin{aligned}
\widetilde{\mathbf{h}} &= \left(\mathbf{R}_{\widetilde{\mathbf{x}}} + \delta \mathbf{I}_{2L}\right)^{-1} \mathbf{R}_{\widetilde{\mathbf{x}}} \widetilde{\mathbf{h}}_{\mathrm{t}} \\
&= \mathbf{Q} \left(\boldsymbol{\Lambda} + \delta \mathbf{I}_{2L}\right)^{-1} \mathbf{Q}^H \mathbf{Q} \boldsymbol{\Lambda} \mathbf{Q}^H \widetilde{\mathbf{h}}_{\mathrm{t}} \\
&= \left(\sum_{l=0}^{2L-1} \frac{\lambda_l}{\lambda_l + \delta} \mathbf{q}_l \mathbf{q}_l^H \right) \widetilde{\mathbf{h}}_{\mathrm{t}} \\
&= \left(\sum_{l=1}^{2L-1} \frac{\lambda_l}{\lambda_l + \delta} \mathbf{q}_l \mathbf{q}_l^H \right) \widetilde{\mathbf{h}}_{\mathrm{t}} \\
&\approx \left(\sum_{l=1}^{2L-1} \mathbf{q}_l \mathbf{q}_l^H \right) \widetilde{\mathbf{h}}_{\mathrm{t}} \\
&\approx \left(\sum_{l=0}^{2L-1} \mathbf{q}_l \mathbf{q}_l^H - \mathbf{q}_0 \mathbf{q}_0^H \right) \widetilde{\mathbf{h}}_{\mathrm{t}} \\
&\approx \widetilde{\mathbf{h}}_{\mathrm{t}} - \left(\mathbf{q}_0^H \widetilde{\mathbf{h}}_{\mathrm{t}} \right) \mathbf{q}_0 \\
&\approx \widetilde{\mathbf{h}}_{\mathrm{t}} + \beta_{\mathrm{f}} \widetilde{\mathbf{g}}.
\end{aligned} \tag{3.73}$$

However, we know that $\widetilde{\mathbf{h}}_{\mathrm{t}}$ is usually sparse. Therefore, to promote sparsity in the solution, it is more appropriate to use the ℓ_1 norm [10], [11].

Taking the gradient of $J_{\mathrm{reg},1}\left(\widetilde{\mathbf{h}}\right)$ with respect to $\widetilde{\mathbf{h}}^H$ and equating the result to zero, we find that [10]

$$-\mathbf{p}_{\widetilde{\mathbf{x}}d} + \mathbf{R}_{\widetilde{\mathbf{x}}} \widetilde{\mathbf{h}} + \frac{\delta}{2} \boldsymbol{\Upsilon} \widetilde{\mathbf{h}} = \mathbf{0}, \tag{3.74}$$

where

$$\boldsymbol{\Upsilon} = \mathrm{diag} \left[\left| \widetilde{h}_0 \right|^{-1}, \left| \widetilde{h}_1 \right|^{-1}, \ldots, \left| \widetilde{h}_{2L-1} \right|^{-1} \right] \tag{3.75}$$

is a diagonal matrix and $\widetilde{h}_l$, $l = 0, 1, \ldots, 2L-1$ are the components of $\widetilde{\mathbf{h}}$. Equation (3.74) can be rewritten as

$$\begin{aligned}
\widetilde{\mathbf{h}} &= \left(\mathbf{R}_{\widetilde{\mathbf{x}}} + \frac{\delta}{2} \boldsymbol{\Upsilon} \right)^{-1} \mathbf{p}_{\widetilde{\mathbf{x}}d} \\
&= \boldsymbol{\Upsilon}^{-1/2} \left(\boldsymbol{\Upsilon}^{-1/2} \mathbf{R}_{\widetilde{\mathbf{x}}} \boldsymbol{\Upsilon}^{-1/2} + \frac{\delta}{2} \mathbf{I}_{2L} \right)^{-1} \boldsymbol{\Upsilon}^{-1/2} \mathbf{p}_{\widetilde{\mathbf{x}}d} \\
&= \mathbf{G} \left(\mathbf{G} \mathbf{R}_{\widetilde{\mathbf{x}}} \mathbf{G} + \frac{\delta}{2} \mathbf{I}_{2L} \right)^{-1} \mathbf{G} \mathbf{p}_{\widetilde{\mathbf{x}}d},
\end{aligned} \tag{3.76}$$

where

$$\mathbf{G} = \boldsymbol{\Upsilon}^{-1/2}$$

$$= \text{diag}\left[\sqrt{\left|\widetilde{h}_0\right|}, \sqrt{\left|\widetilde{h}_1\right|}, \ldots, \sqrt{\left|\widetilde{h}_{2L-1}\right|}\right]. \qquad (3.77)$$

As suggested in [10], (3.75) can be solved iteratively as follows:

$$\widetilde{\mathbf{h}}(n+1) = \mathbf{G}(n)\left[\mathbf{G}(n)\mathbf{R}_{\widetilde{\mathbf{x}}}\mathbf{G}(n) + \frac{\delta}{2}\mathbf{I}_{2L}\right]^{-1}\mathbf{G}(n)\mathbf{p}_{\widetilde{\mathbf{x}}d}, \qquad (3.78)$$

where

$$\mathbf{G}(n) = \text{diag}\left[\sqrt{\left|\widetilde{h}_0(n)\right|}, \sqrt{\left|\widetilde{h}_1(n)\right|}, \ldots, \sqrt{\left|\widetilde{h}_{2L-1}(n)\right|}\right]. \qquad (3.79)$$

We deduce that, if $\widetilde{\mathbf{h}}_{\text{t}}$ has L_0 elements equal to 0, the linear system of $2L$ equations in (3.76) is reduced to a linear system of $2L - L_0$ equations. In other words, the $2L \times 2L$ matrix $\mathbf{G}\mathbf{R}_{\widetilde{\mathbf{x}}}\mathbf{G}$ can be reduced to a $(2L - L_0) \times (2L - L_0)$ matrix. As a consequence, the latter matrix is not only better conditioned but it may become full rank. So one interesting way to help solving the SAEC problem is to take into account the sparseness of the acoustic impulse responses in the adaptive algorithms.

Another equivalent way to write (3.71) is

$$\begin{aligned} J_{\text{reg},\boldsymbol{\Upsilon}}\left(\widetilde{\mathbf{h}}\right) &= J\left(\widetilde{\mathbf{h}}\right) + \frac{\delta}{2}\left\|\widetilde{\mathbf{h}}\right\|_{\boldsymbol{\Upsilon}}^2 \\ &= E\left[|e(n)|^2\right] + \frac{\delta}{2}\widetilde{\mathbf{h}}^H\boldsymbol{\Upsilon}\widetilde{\mathbf{h}}, \end{aligned} \qquad (3.80)$$

where $\boldsymbol{\Upsilon}$ is a Hermitian positive-definite matrix. The filter obtained from minimizing $J_{\text{reg},\boldsymbol{\Upsilon}}\left(\widetilde{\mathbf{h}}\right)$ is exactly the one given in (3.76) but now $\mathbf{G}$ can be any Hermitian positive-definite matrix. With a judicious choice of the elements of $\mathbf{G}$ and making them dependent on the coefficients of the acoustic impulse response, we can promote sparsity. The form given in (3.80) is convenient to use in adaptive filters.

References

1. N. Wiener, *Extrapolation, Interpolation, and Smoothing of Stationary Time Series.* New York: John Wiley & Sons, 1949.
2. S. Haykin, *Adaptive Filter Theory.* Fourth Edition, Upper Saddle River, NJ: Prentice-Hall, 2002.
3. M. M. Sondhi, D. R. Morgan, and J. L. Hall, "Stereophonic acoustic echo cancellation–An overview of the fundamental problem," *IEEE Signal Process. Lett.*, vol. 2, pp. 148–151, Aug. 1995.
4. J. Benesty, P. Duhamel, and Y. Grenier, "Multi-channel adaptive filtering applied to multi-channel acoustic echo cancellation," in *Proc. EUSIPCO*, 1996, pp. 1405–1408.

5. J. Benesty, D. R. Morgan, and M. M. Sondhi, "A better understanding and an improved solution to the specific problems of stereophonic acoustic echo cancellation," *IEEE Trans. Speech, Audio Process.*, vol. 6, pp. 156–165, Mar. 1998.
6. B. Widrow, "Adaptive filters," in *Aspects of Network and System Theory*, R. E. Kalman and N. DeClaris, Eds., New York: Holt, Rinehart and Winston, 1970.
7. J. Benesty, T. Gänsler, D. R. Morgan, M. M. Sondhi, and S. L. Gay, *Advances in Network and Acoustic Echo Cancellation.* Berlin, Germany: Springer-Verlag, 2001.
8. B. C. J. Moore, *An Introduction to the Psychology of Hearing.* London, UK: Academic Press, 1989.
9. D. R. Morgan, J. L. Hall, and J. Benesty, "Investigation of several types of nonlinearities for use in stereo acoustic echo cancellation," *IEEE Trans. Speech, Audio Process.*, vol. 9, pp. 686–696, Sept. 2001.
10. B. D. Rao and B. Song, "Adaptive filtering algorithms for promoting sparsity," in *Proc. IEEE ICASSP*, 2003, pp. VI-361–VI-364.
11. M. S. Asif and J. Romberg, "Dynamic updating for ℓ_1 minimization," *IEEE J. Sel. Topics Signal Process.*, vol. 4, pp. 421–434, Apr. 2010.

Chapter 4
A Class of Stochastic Adaptive Filters

In this chapter, we derive, study, and analyze a class of stochastic adaptive filters for SAEC with the WL model. All developed algorithms try to converge to the optimal Wiener filter. We start with the classical stochastic gradient algorithm, which is a good approximation of the deterministic algorithm studied in the previous chapter.

4.1 Least-Mean-Square (LMS) Algorithm

The least-mean-square (LMS) or stochastic gradient algorithm, invented by Widrow and Hoff in the late 50's [1], is certainly the most popular algorithm that can be found in the literature of adaptive filters. Its popularity is, perhaps, due to the fact that it is easy to understand, convenient to implement, and robust in many respects.

The conventional way to derive the stereo stochastic gradient algorithm is by approximating the deterministic algorithm. Indeed, in practice, the two quantities $\mathbf{p}_{\widetilde{\mathbf{x}}d}$ and $\mathbf{R}_{\widetilde{\mathbf{x}}}$ are, in general, not known. If we take their instantaneous estimates, i.e.,

$$\widehat{\mathbf{p}}_{\widetilde{\mathbf{x}}d}(n) = \widetilde{\mathbf{x}}(n)d^*(n), \tag{4.1}$$

$$\widehat{\mathbf{R}}_{\widetilde{\mathbf{x}}}(n) = \widetilde{\mathbf{x}}(n)\widetilde{\mathbf{x}}^H(n), \tag{4.2}$$

and replace them in the steepest-descent algorithm [eq. (3.40)], we get

$$\begin{aligned}\widetilde{\mathbf{h}}(n) &= \widetilde{\mathbf{h}}(n-1) + \mu\left[\widehat{\mathbf{p}}_{\widetilde{\mathbf{x}}d}(n) - \widehat{\mathbf{R}}_{\widetilde{\mathbf{x}}}(n)\widetilde{\mathbf{h}}(n-1)\right] \\ &= \widetilde{\mathbf{h}}(n-1) + \mu\widetilde{\mathbf{x}}(n)\left[d^*(n) - \widetilde{\mathbf{x}}^H(n)\widetilde{\mathbf{h}}(n-1)\right]. \end{aligned} \tag{4.3}$$

This simple recursion is the LMS algorithm. Contrary to the deterministic algorithm, the LMS weight vector $\widetilde{\mathbf{h}}(n)$ is now a random vector. The following

J. Benesty et al.: A Perspective on Stereophonic Acoustic Echo Cancellation, STSP 4, pp. 29–48.
springerlink.com

three equations summarize this algorithm [1], [2]:

$$\widehat{y}(n) = \widetilde{\mathbf{h}}^H(n-1)\widetilde{\mathbf{x}}(n), \text{ filter output,} \tag{4.4}$$

$$e(n) = d(n) - \widehat{y}(n), \text{ error signal,} \tag{4.5}$$

$$\widetilde{\mathbf{h}}(n) = \widetilde{\mathbf{h}}(n-1) + \mu\widetilde{\mathbf{x}}(n)e^*(n), \text{ adaptation,} \tag{4.6}$$

which requires around $4L$ complex additions and $4L$ complex multiplications at each iteration.

4.2 Performance of the LMS Algorithm

In this section, we study the convergence, the transient behavior, and the steady-state behavior of the LMS algorithm.

Using the misalignment vector, $\mathbf{m}(n) = \widetilde{\mathbf{h}}_t - \widetilde{\mathbf{h}}(n)$, the LMS update equation becomes

$$\mathbf{m}(n) = \left[\mathbf{I}_{2L} - \mu\widetilde{\mathbf{x}}(n)\widetilde{\mathbf{x}}^H(n)\right]\mathbf{m}(n-1) - \mu\widetilde{\mathbf{x}}(n)v^*(n). \tag{4.7}$$

Taking expectation on both sides of (4.7), we get

$$E\left[\mathbf{m}(n)\right] = E\left\{\left[\mathbf{I}_{2L} - \mu\widetilde{\mathbf{x}}(n)\widetilde{\mathbf{x}}^H(n)\right]\mathbf{m}(n-1)\right\}. \tag{4.8}$$

Now, assuming that μ is small, (4.8) can be rewritten as

$$\begin{aligned} E\left[\mathbf{m}(n)\right] &= E\left\{\left[\mathbf{I}_{2L} - \mu\widetilde{\mathbf{x}}(n)\widetilde{\mathbf{x}}^H(n)\right]\right\}E\left[\mathbf{m}(n-1)\right] \\ &= \left[\mathbf{I}_{2L} - \mu\mathbf{R}_{\widetilde{\mathbf{x}}}\right]E\left[\mathbf{m}(n-1)\right]. \end{aligned} \tag{4.9}$$

With the eigendecomposition of $\mathbf{R}_{\widetilde{\mathbf{x}}}$, we can express the previous equation as

$$\begin{aligned} E\left[\mathbf{t}(n)\right] &= \left[\mathbf{I}_{2L} - \mu\mathbf{\Lambda}\right]E\left[\mathbf{t}(n-1)\right] \\ &= \left[\mathbf{I}_{2L} - \mu\mathbf{\Lambda}\right]^n E\left[\mathbf{t}(0)\right], \end{aligned} \tag{4.10}$$

where $\mathbf{t}(n) = \mathbf{Q}^H\mathbf{m}(n)$ (see Chapter 3).

We say that the LMS converges in the mean to the true impulse response if

$$\lim_{n\to\infty} E\left[\mathbf{t}(n)\right] = \mathbf{0}. \tag{4.11}$$

In this case,

$$\lim_{n\to\infty} E\left[\widetilde{\mathbf{h}}(n)\right] = \widetilde{\mathbf{h}}_t. \tag{4.12}$$

We see from (4.10) that a necessary condition for the LMS to converge in the mean to the desired solution is that

$$0 < \mu < \frac{2}{\lambda_{\max}}. \tag{4.13}$$

This condition is not sufficient since for the SAEC problem (i.e., when $\lambda_0 = 0$), we have

$$E\left[t_0(\infty)\right] = t_0(0) = \mathbf{q}_0^H \widetilde{\mathbf{h}}_{\mathrm{t}}, \tag{4.14}$$
$$E\left[t_l(\infty)\right] = 0, \ \forall l \neq 0, \tag{4.15}$$

assuming that the LMS is initialized with $\widetilde{\mathbf{h}}(0) = \mathbf{0}$. As a result,

$$\begin{aligned} E\left[\widetilde{\mathbf{h}}(\infty)\right] &= \widetilde{\mathbf{h}}_{\mathrm{t}} - \mathbf{Q}E\left[\mathbf{t}(\infty)\right] \\ &= \widetilde{\mathbf{h}}_{\mathrm{t}} - \left(\mathbf{q}_0^H \widetilde{\mathbf{h}}_{\mathrm{t}}\right) \mathbf{q}_0 \\ &= \widetilde{\mathbf{h}}_{\mathrm{t}} + \beta_{\mathrm{f}} \widetilde{\mathbf{g}}. \end{aligned} \tag{4.16}$$

Although, the LMS is convergent in the mean with the condition in (4.13), it is biased and $E\left[\widetilde{\mathbf{h}}(\infty)\right]$ may be far away from the desired solution, $\widetilde{\mathbf{h}}_{\mathrm{t}}$. In this difficult situation, there is not much we can do unless we distort the input signal, $x(n)$, such that $\lambda_l > 0, \ \forall l$.

To study the behavior of the LMS in the mean square, we need to evaluate the covariance matrix of $\mathbf{t}(n)$, i.e.,

$$\mathbf{R}_{\mathbf{t}}(n) = E\left[\mathbf{t}(n)\mathbf{t}^H(n)\right]. \tag{4.17}$$

Assuming that μ is small and $\mathbf{m}(n-1)$ and $v(n)$ are independent (which is true for a white Gaussian noise), we obtain

$$\begin{aligned} \mathbf{R}_{\mathbf{t}}(n) = \mathbf{R}_{\mathbf{t}}(n-1) - \mu \mathbf{R}_{\mathbf{t}}(n-1)\boldsymbol{\Lambda} - \mu \boldsymbol{\Lambda} \mathbf{R}_{\mathbf{t}}(n-1) \\ + \mu^2 \boldsymbol{\Lambda} \mathbf{R}_{\mathbf{t}}(n-1)\boldsymbol{\Lambda} + \mu^2 \sigma_v^2 \boldsymbol{\Lambda}. \end{aligned} \tag{4.18}$$

Since we are only interested in the diagonal elements of $\mathbf{R}_{\mathbf{t}}(n)$, we deduce that

$$\begin{aligned} E\left[\left|t_l(n)\right|^2\right] &= \left(1 - \mu\lambda_l\right)^2 E\left[\left|t_l(n-1)\right|^2\right] + \mu^2 \lambda_l \sigma_v^2 \\ &= \left(1 - \mu\lambda_l\right)^{2n} E\left[\left|t_l(0)\right|^2\right] + \mu^2 \lambda_l \sigma_v^2 \sum_{i=0}^{n-1} \left(1 - \mu\lambda_l\right)^{2i} \\ &= \frac{\mu \sigma_v^2}{2 - \mu\lambda_l} + \left(1 - \mu\lambda_l\right)^{2n} \left\{ E\left[\left|t_l(0)\right|^2\right] - \frac{\mu\sigma_v^2}{2 - \mu\lambda_l} \right\}. \end{aligned} \tag{4.19}$$

Let us define the misalignment

$$\begin{aligned}\mathcal{H}(n) &= E\left[\mathbf{m}^H(n)\mathbf{m}(n)\right] \\ &= E\left[\mathbf{t}^H(n)\mathbf{t}(n)\right] \\ &= E\left[\left\|\widetilde{\mathbf{h}}_{\mathrm{t}} - \widetilde{\mathbf{h}}(n)\right\|_2^2\right] \\ &= \sum_{l=0}^{2L-1} E\left[|t_l(n)|^2\right].\end{aligned} \tag{4.20}$$

We say that the LMS converges in the mean square if

$$\lim_{n\to\infty} \mathcal{H}(n) < \infty. \tag{4.21}$$

A necessary and sufficient condition for that is

$$0 < \mu < \frac{2}{\lambda_{\max}}, \tag{4.22}$$

which is identical to the condition on the convergence of the LMS in the mean and the convergence of the deterministic algorithm.

With the previous condition and with $\mathbf{R}_{\widetilde{\mathbf{x}}} > 0$, we obtain

$$\mathcal{H}(\infty) = \sum_{l=0}^{2L-1} \frac{\mu\sigma_v^2}{2 - \mu\lambda_l}. \tag{4.23}$$

We see how the value of $\mathcal{H}(\infty)$ depends on the variance of the noise, σ_v^2, and the step-size parameter, μ. The smaller the value of μ or σ_v^2, the better the misalignment. But with $\lambda_0 = 0$, the steady-state value of the misalignment is

$$\mathcal{H}(\infty) = \left\|\mathbf{q}_0^H\widetilde{\mathbf{h}}_{\mathrm{t}}\right\|_2^2 + \sum_{l=1}^{2L-1} \frac{\mu\sigma_v^2}{2 - \mu\lambda_l} \geq \left\|\mathbf{q}_0^H\widetilde{\mathbf{h}}_{\mathrm{t}}\right\|_2^2. \tag{4.24}$$

We observe now how the misalignment is lower bounded no matter the values of μ and/or σ_v^2.

Assuming that $v(n)$, $\widetilde{\mathbf{x}}(n)$, and $\mathbf{m}(n-1)$ are mutually independent, the MSE produced by the LMS is given by

$$\begin{aligned}J(n) &= E\left[|e(n)|^2\right] \\ &= \sigma_v^2 + E\left[\mathbf{m}^H(n-1)\widetilde{\mathbf{x}}(n)\widetilde{\mathbf{x}}^H(n)\mathbf{m}(n-1)\right] \\ &= \sigma_v^2 + \mathrm{tr}\left[\mathbf{R}_{\widetilde{\mathbf{x}}}\mathbf{R}_{\mathbf{m}}(n-1)\right] \\ &= \sigma_v^2 + \mathrm{tr}\left[\mathbf{\Lambda}\mathbf{R}_{\mathbf{t}}(n-1)\right] \\ &= \sigma_v^2 + \sum_{l=0}^{2L-1} \lambda_l E\left[|t_l(n-1)|^2\right],\end{aligned} \tag{4.25}$$

where $\mathrm{tr}[\cdot]$ denotes the trace of a square matrix and $\mathbf{R}_{\mathbf{m}}(n-1) = E\left[\mathbf{m}(n-1)\mathbf{m}^H(n-1)\right]$. It is interesting to observe that the stereo echo is always cancelled even when $\mathbf{R}_{\widetilde{\mathbf{x}}}$ is not full rank; but it is even more interesting to see that in this case we have more echo cancellation than when $\mathbf{R}_{\widetilde{\mathbf{x}}} > 0$. We deduce that the steady-state value of the MSE is

$$\begin{aligned} J(\infty) &= \sigma_v^2 + \mu\sigma_v^2 \sum_{l=0}^{2L-1} \frac{\lambda_l}{2-\mu\lambda_l} \\ &\approx \sigma_v^2 \left[1 + \frac{\mu}{2}\mathrm{tr}\left(\mathbf{R}_{\widetilde{\mathbf{x}}}\right)\right], \ \mu \text{ small.} \end{aligned} \tag{4.26}$$

We define the excess MSE (EMSE) as the difference between the MSE of the LMS and the minimum MSE, $J_{\min}$, of the Wiener filter, i.e.,

$$\begin{aligned} J_{\mathrm{ex}}(n) &= J(n) - J_{\min} \\ &= \sum_{l=0}^{2L-1} \lambda_l E\left[\left|t_l(n-1)\right|^2\right]. \end{aligned} \tag{4.27}$$

As a result, the steady-state value of the EMSE is

$$\begin{aligned} J_{\mathrm{ex}}(\infty) &= \mu\sigma_v^2 \sum_{l=0}^{2L-1} \frac{\lambda_l}{2-\mu\lambda_l} \\ &\approx \frac{\mu\sigma_v^2}{2}\mathrm{tr}\left(\mathbf{R}_{\widetilde{\mathbf{x}}}\right), \ \mu \text{ small.} \end{aligned} \tag{4.28}$$

The misadjustment is defined as [3]

$$\mathcal{M} = \frac{J_{\mathrm{ex}}(\infty)}{J_{\min}}. \tag{4.29}$$

Hence,

$$\begin{aligned} \mathcal{M} &= \mu \sum_{l=0}^{2L-1} \frac{\lambda_l}{2-\mu\lambda_l} \\ &\approx \frac{\mu}{2}\mathrm{tr}\left(\mathbf{R}_{\widetilde{\mathbf{x}}}\right), \ \mu \text{ small.} \end{aligned} \tag{4.30}$$

The misadjustment measures how well $\mu E\left[\widetilde{\mathbf{x}}(n)e^*(n)\right]$ is approximated by $\mu\widetilde{\mathbf{x}}(n)e^*(n)$ at the steady state.

4.3 Normalized LMS (NLMS) Algorithm

Another way to find the stability condition of the LMS is from the inequality

$$|\epsilon(n)| < |e(n)|, \tag{4.31}$$

where

$$\epsilon(n) = d(n) - \widetilde{\mathbf{h}}^H(n)\widetilde{\mathbf{x}}(n) \tag{4.32}$$

is the a posteriori error signal, computed after the filter is updated. This intuitively makes sense since $\epsilon(n)$ contains more meaningful information than $e(n)$.

This condition is necessary for the LMS to converge to the impulse response of the system but not sufficient. However, it is very useful for finding the bounds for the step size μ.

Plugging (4.6) in (4.32) and using the condition (4.31), we find

$$0 < \mu < \frac{2}{\widetilde{\mathbf{x}}^H(n)\widetilde{\mathbf{x}}(n)}. \tag{4.33}$$

For a large L and a stationary input signal, $\widetilde{\mathbf{x}}^H(n)\widetilde{\mathbf{x}}(n) \approx 2L\sigma_x^2 = \mathrm{tr}\left(\mathbf{R}_{\widetilde{\mathbf{x}}}\right)$. On the other hand, $\mathrm{tr}\left(\mathbf{R}_{\widetilde{\mathbf{x}}}\right) = \sum_{l=0}^{2L-1} \lambda_l$ and this implies that $\mathrm{tr}\left(\mathbf{R}_{\widetilde{\mathbf{x}}}\right) \geq \lambda_{\max}$. Hence,

$$0 < \mu < \frac{2}{\widetilde{\mathbf{x}}^H(n)\widetilde{\mathbf{x}}(n)} \leq \frac{2}{\lambda_{\max}}. \tag{4.34}$$

If we now introduce the normalized step size α $(0 < \alpha < 2)$, the step size of the LMS varies with time as follows:

$$\mu(n) = \frac{\alpha}{\widetilde{\mathbf{x}}^H(n)\widetilde{\mathbf{x}}(n)}, \tag{4.35}$$

and the LMS becomes the normalized LMS (NLMS):

$$\widetilde{\mathbf{h}}(n) = \widetilde{\mathbf{h}}(n-1) + \frac{\alpha\widetilde{\mathbf{x}}(n)e^*(n)}{\widetilde{\mathbf{x}}^H(n)\widetilde{\mathbf{x}}(n)}. \tag{4.36}$$

This algorithm is extremely helpful in practice, especially with nonstationary signals, since $\mu(n)$ can adjust itself at each new iteration. As a consequence, the NLMS algorithm converges and tracks better than the standard LMS algorithm [4].

A very useful way to derive the NLMS algorithm in the particular case of $\alpha = 1$ is by minimizing $\left\|\widetilde{\mathbf{h}}(n) - \widetilde{\mathbf{h}}(n-1)\right\|_2^2$ with the constraint that $\epsilon(n) = 0$. This shows again the importance of the a posteriori error signal in adaptive filters.

We can deduce that for $\mathbf{R}_{\widetilde{\mathbf{x}}} > 0$, we have

$$\mathcal{H}(\infty) \approx \frac{\alpha}{2} \cdot \frac{\sigma_v^2}{\sigma_x^2} \tag{4.37}$$

and for $\lambda_0 = 0$, we have

$$\mathcal{H}(\infty) \approx \left\| \mathbf{q}_0^H \widetilde{\mathbf{h}}_\mathrm{t} \right\|_2^2 + \frac{\alpha}{2} \cdot \frac{\sigma_v^2}{\sigma_x^2}. \tag{4.38}$$

We also deduce the steady-state values of the MSE and EMSE, and the misadjustment for the NLMS:

$$J(\infty) \approx \sigma_v^2 \left(1 + \frac{\alpha}{2}\right), \tag{4.39}$$

$$J_\mathrm{ex}(\infty) \approx \frac{\alpha}{2} \cdot \sigma_v^2, \tag{4.40}$$

$$\mathcal{M} \approx \frac{\alpha}{2}. \tag{4.41}$$

Finally, to end this section, we give the regularized version of the NLMS algorithm:

$$\widetilde{\mathbf{h}}(n) = \widetilde{\mathbf{h}}(n-1) + \frac{\alpha \widetilde{\mathbf{x}}(n) e^*(n)}{\widetilde{\mathbf{x}}^H(n)\widetilde{\mathbf{x}}(n) + \delta}, \tag{4.42}$$

where $\delta \geq 0$ is the regularization parameter. Regularization is important in practice when the input signal is ill-conditioned and, especially, in the presence of noise.

4.4 Interpretation of the NLMS Algorithm

In this section, we assume that $\alpha = 1$ and $\delta = 0$. In this case, it is easy to see that the update equation of the NLMS can be rewritten as [5]

$$\widetilde{\mathbf{h}}(n) = \mathbf{P}(n)\widetilde{\mathbf{h}}(n-1) + \overleftarrow{\mathbf{h}}(n), \tag{4.43}$$

where

$$\mathbf{P}(n) = \mathbf{I}_{2L} - \frac{\widetilde{\mathbf{x}}(n)\widetilde{\mathbf{x}}^H(n)}{\widetilde{\mathbf{x}}^H(n)\widetilde{\mathbf{x}}(n)} \tag{4.44}$$

is an orthogonal projection matrix whose rank is equal to $2L - 1$ and

$$\overleftarrow{\mathbf{h}}(n) = \frac{\widetilde{\mathbf{x}}(n) d^*(n)}{\widetilde{\mathbf{x}}^H(n)\widetilde{\mathbf{x}}(n)} \tag{4.45}$$

is the correction component of the NLMS algorithm. We observe that $\widetilde{\mathbf{h}}(n)$ is the sum of two vectors that are orthogonal, i.e.,

$$\left[\mathbf{P}(n)\widetilde{\mathbf{h}}(n-1)\right]^H \overleftarrow{\mathbf{h}}(n) = 0. \tag{4.46}$$

Therefore, $\overleftarrow{\mathbf{h}}(n)$ belongs to the nullspace of $\mathbf{P}(n)$, i.e.,

$$\mathbf{P}(n)\overleftarrow{\mathbf{h}}(n) = \mathbf{0}. \tag{4.47}$$

Furthermore $\widetilde{\mathbf{h}}(n-1)$, in the space formed from the complementary (range and null) subspaces of $\mathbf{P}(n)$, can be decomposed uniquely as

$$\widetilde{\mathbf{h}}(n-1) = \widetilde{\mathbf{h}}_{\|}(n-1) + \widetilde{\mathbf{h}}_{\perp}(n-1), \tag{4.48}$$

where

$$\mathbf{P}(n)\widetilde{\mathbf{h}}(n-1) = \widetilde{\mathbf{h}}_{\|}(n-1), \tag{4.49}$$

$$\mathbf{P}(n)\widetilde{\mathbf{h}}_{\perp}(n-1) = \mathbf{0}. \tag{4.50}$$

As a result, (4.43) simplifies to

$$\widetilde{\mathbf{h}}(n) = \widetilde{\mathbf{h}}_{\|}(n-1) + \overleftarrow{\mathbf{h}}(n), \tag{4.51}$$

since the component $\widetilde{\mathbf{h}}_{\perp}(n-1)$ does not affect the update equation. Clearly, we have $\widetilde{\mathbf{h}}_{\|}^H(n-1)\overleftarrow{\mathbf{h}}(n) = 0$ and $\widetilde{\mathbf{x}}^H(n)\widetilde{\mathbf{h}}(n) = \widetilde{\mathbf{x}}^H(n)\overleftarrow{\mathbf{h}}(n) = d^*(n)$. According to (4.48) and (4.49),

$$\widetilde{\mathbf{h}}_{\perp}(n-1) = \left[\mathbf{I}_{2L} - \mathbf{P}(n)\right]\widetilde{\mathbf{h}}(n-1). \tag{4.52}$$

Then, using (4.44) in (4.52), we get

$$\widetilde{\mathbf{h}}_{\perp}(n-1) = \frac{\widetilde{\mathbf{x}}(n)\widehat{y}^*(n)}{\widetilde{\mathbf{x}}^H(n)\widetilde{\mathbf{x}}(n)}, \tag{4.53}$$

which is interesting to compare to $\overleftarrow{\mathbf{h}}(n)$. Both vectors are obviously in the nullspace of $\mathbf{P}(n)$. Another way, then, to write the error signal is

$$e(n) = \left[\overleftarrow{\mathbf{h}}(n) - \widetilde{\mathbf{h}}_{\perp}(n-1)\right]^H \widetilde{\mathbf{x}}(n), \tag{4.54}$$

which is not affected by $\widetilde{\mathbf{h}}_{\|}(n-1)$.

The most remarkable part in the decomposition of $\widetilde{\mathbf{h}}(n)$ in (4.51) is that $\overleftarrow{\mathbf{h}}(n)$ is, in fact, the minimum ℓ_2-norm solution of the linear system of one equation $d(n) = \overleftarrow{\mathbf{h}}^H(n)\widetilde{\mathbf{x}}(n)$. Indeed, by

$$\min_{\overleftarrow{\mathbf{h}}(n)} \left\| \overleftarrow{\mathbf{h}}(n) \right\|_2^2 \quad \text{subject to} \quad d(n) = \overleftarrow{\mathbf{h}}^H(n)\widetilde{\mathbf{x}}(n), \tag{4.55}$$

we find (4.45). And, of course, $\widetilde{\mathbf{h}}(n)$ is also a solution of $d(n) = \overleftarrow{\mathbf{h}}^H(n)\widetilde{\mathbf{x}}(n)$. Therefore, $\widetilde{\mathbf{h}}_{\|}(n-1)$ can be seen as a good initialization of the adaptive filter since the minimum ℓ_2-norm solution is not the optimal solution.

The interpretation [5] of the NLMS algorithm is the following. First, we find the minimum ℓ_2-norm solution of $d(n) = \overleftarrow{\mathbf{h}}^H(n)\widetilde{\mathbf{x}}(n)$, which is denoted by $\overleftarrow{\mathbf{h}}(n)$. This step is the most important one since it determines the adaptation with the most recent information. Second, we form an (orthogonal) projection matrix, $\mathbf{P}(n)$, of rank $2L-1$ for which $\overleftarrow{\mathbf{h}}(n)$ spans its nullspace. Third, from $\widetilde{\mathbf{h}}(n-1)$ we find the initialization vector, $\widetilde{\mathbf{h}}_{\|}(n-1)$, that is in the range of $\mathbf{P}(n)$. Finally, we add $\overleftarrow{\mathbf{h}}(n)$ and $\widetilde{\mathbf{h}}_{\|}(n-1)$ to find the new update equation resulting in the NLMS.

4.5 Regularization of the NLMS Algorithm

The choice of the value of the regularization parameter, δ, is as important as the choice of the value of the normalized step-size parameter, α, in the NLMS algorithm. In practice, we often take

$$\delta = \beta\sigma_x^2, \tag{4.56}$$

where β is the normalized (with respect to the variance of the input signal) regularization parameter. Next, we show how to find β (see also [6]).

It can be checked that the update equation of the NLMS can be rewritten as

$$\widetilde{\mathbf{h}}(n) = \mathbf{P}'(n)\widetilde{\mathbf{h}}(n-1) + \alpha\overleftarrow{\mathbf{h}}'(n), \tag{4.57}$$

where now

$$\mathbf{P}'(n) = \mathbf{I}_{2L} - \alpha\frac{\widetilde{\mathbf{x}}(n)\widetilde{\mathbf{x}}^H(n)}{\widetilde{\mathbf{x}}^H(n)\widetilde{\mathbf{x}}(n) + \delta} \tag{4.58}$$

and

$$\overleftarrow{\mathbf{h}}'(n) = \frac{\widetilde{\mathbf{x}}(n)d^*(n)}{\widetilde{\mathbf{x}}^H(n)\widetilde{\mathbf{x}}(n) + \delta}, \tag{4.59}$$

which depends on the new observation $d(n)$. Note that $\mathbf{P}'(n)$ does not depend on the noise signal. Basically, (4.57) is decomposed into two components: one free of noise and a noisy one. Therefore, as far as regularization is concerned, only the noisy component is of interest.

In fact, $\overleftarrow{\mathbf{h}}'(n)$ is one of the possible solutions of the undetermined linear system of one equation $d(n) = \overleftarrow{\mathbf{h}}'^H(n)\widetilde{\mathbf{x}}(n)$. Clearly, this solution is not the optimal one. The regularized version of the minimum ℓ_2-norm solution of that linear system is obtained by solving

$$\min_{\overleftarrow{\mathbf{h}}'(n)} \left[\left| d(n) - \overleftarrow{\mathbf{h}}'^H(n)\widetilde{\mathbf{x}}(n) \right|^2 + \delta \left\| \overleftarrow{\mathbf{h}}'(n) \right\|_2^2 \right]. \quad (4.60)$$

So the other vector $\mathbf{P}'(n)\widetilde{\mathbf{h}}(n-1)$ in (4.57) can be seen as a good initialization of the adaptive filter as explained in the previous section.

The question now is how to find δ?

Since $\overleftarrow{e}(n) = d(n) - \overleftarrow{\mathbf{h}}'^H(n)\widetilde{\mathbf{x}}(n)$ is the error signal between the desired signal and the estimated signal obtained from the filter optimized in (4.60), we should find δ in such a way the the expected value of $|\overleftarrow{e}(n)|^2$ is equal to the variance of the noise, i.e.,

$$E\left[|\overleftarrow{e}(n)|^2 \right] = \sigma_v^2. \quad (4.61)$$

This is reasonable if we want to attenuate the effects of the noise in the estimator $\overleftarrow{\mathbf{h}}'(n)$.

To derive the optimal δ according to (4.61), we assume that $L \gg 1$ and $x(n)$ is stationary. As a result,

$$\widetilde{\mathbf{x}}^H(n)\widetilde{\mathbf{x}}(n) \approx 2L\sigma_x^2. \quad (4.62)$$

Developing (4.61) and using (4.62), we easily derive the quadratic equation:

$$\delta^2 - 2\frac{2L\sigma_x^2}{\text{SENR}}\delta - \frac{\left(2L\sigma_x^2\right)^2}{\text{SENR}} = 0, \quad (4.63)$$

from which we deduce the obvious solution

$$\begin{aligned} \delta &= \frac{2L\left(1 + \sqrt{1 + \text{SENR}}\right)}{\text{SENR}}\sigma_x^2 \\ &= \beta_{\text{NLMS}}\sigma_x^2, \end{aligned} \quad (4.64)$$

where

$$\beta_{\text{NLMS}} = \frac{2L\left(1 + \sqrt{1 + \text{SENR}}\right)}{\text{SENR}} \quad (4.65)$$

is the normalized regularization parameter of the NLMS.

We see that δ depends on three terms: the length $2L$ of the adaptive filter, the variance, σ_x^2, of the input signal, and the SENR. In acoustic echo cancellation, the first two terms ($2L$ and σ_x^2) are known, while the SENR is often roughly known or can be estimated. Therefore, it is not hard to find a good value of δ for SAEC.

Furthermore, we have

$$\lim_{\text{SENR}\to 0} \delta = \infty, \tag{4.66}$$

$$\lim_{\text{SENR}\to \infty} \delta = 0, \tag{4.67}$$

which is what we desire.

With the proposed regularization parameter, it can be verified that the NLMS can be expressed as

$$\widetilde{\mathbf{h}}(n) = \widetilde{\mathbf{h}}(n-1) + \frac{\alpha}{2L\sigma_x^2} \cdot \frac{\text{SENR}}{1 + \text{SENR} + \sqrt{1+\text{SENR}}} \widetilde{\mathbf{x}}(n) e^*(n). \tag{4.68}$$

We then deduce that the misadjustment is

$$\mathcal{M} \approx \frac{\alpha}{2} \cdot \frac{\text{SENR}}{1 + \text{SENR} + \sqrt{1+\text{SENR}}}, \tag{4.69}$$

which is always smaller than the misadjustment produced by the unregularized NLMS. Furthermore, many simulations show that the proposed regularization technique does not affect the convergence and tracking of the NLMS, while it is very stable with different input signals and levels of the SENR.

4.6 Variable Step-Size NLMS (VSS-NLMS) Algorithm

The stability of the NLMS algorithm is governed by the normalized step-size parameter, α. The choice of this parameter, within the stability conditions, reflects a tradeoff between fast convergence and good tracking ability on one hand, and low misadjustment on the other hand. A value of α equal to 1 will produce the fastest convergence rate but with a large misadjustment. A value of α smaller than 1 will reduce the convergence rate but will also give a lower misadjustment. To meet this conflicting requirement, the step size needs to be controlled. While the formulation of this problem is straightforward, a good and reliable solution is not that easy to find. Many different schemes have been proposed in the last two decades [7], [8], [9], [10], [11], [12], [13], [14]. In this section, we show how to derive, in a very simple and elegant way, a non-parametric variable step-size NLMS algorithm [15].

We can rewrite the a priori and a posteriori error signals as, respectively,

$$e(n) = d(n) - \widetilde{\mathbf{h}}^H(n-1)\widetilde{\mathbf{x}}(n) \tag{4.70}$$

$$= \left[\widetilde{\mathbf{h}}_{\mathrm{t}} - \widetilde{\mathbf{h}}(n-1)\right]^H \widetilde{\mathbf{x}}(n) + v(n),$$

$$\begin{aligned}\epsilon(n) &= d(n) - \widetilde{\mathbf{h}}^H(n)\widetilde{\mathbf{x}}(n) \\ &= \left[\widetilde{\mathbf{h}}_{\mathrm{t}} - \widetilde{\mathbf{h}}(n)\right]^H \widetilde{\mathbf{x}}(n) + v(n).\end{aligned} \tag{4.71}$$

Consider the update equation

$$\widetilde{\mathbf{h}}(n) = \widetilde{\mathbf{h}}(n-1) + \mu(n)\widetilde{\mathbf{x}}(n)e^*(n). \tag{4.72}$$

One reasonable way to derive a $\mu(n)$ that makes (4.72) stable is to cancel the a posteriori error signal (see [16] and references therein). Replacing (4.72) in (4.71) with the requirement $\epsilon(n) = 0$, we easily find assuming $e(n) \neq 0,\ \forall n$, that

$$\mu_{\mathrm{NLMS}}(n) = \frac{1}{\widetilde{\mathbf{x}}^H(n)\widetilde{\mathbf{x}}(n)}. \tag{4.73}$$

Therefore, the obtained algorithm is the classical NLMS.

While the above procedure makes sense in the absence of noise, finding the $\mu(n)$ in the presence of noise, that cancels (4.71) will introduce noise in $\widetilde{\mathbf{h}}(n)$ since $\left[\widetilde{\mathbf{h}}_{\mathrm{t}} - \widetilde{\mathbf{h}}(n)\right]^H \widetilde{\mathbf{x}}(n) = -v(n) \neq 0,\ \forall n$. What we would like, in fact, is to have $\left[\widetilde{\mathbf{h}}_{\mathrm{t}} - \widetilde{\mathbf{h}}(n)\right]^H \widetilde{\mathbf{x}}(n) = 0,\ \forall n$, which implies that $\epsilon(n) = v(n)$. Hence, in this procedure, we wish to find the step-size parameter $\mu(n)$ in such a way that

$$E\left[|\epsilon(n)|^2\right] = \sigma_v^2,\ \forall n. \tag{4.74}$$

Using the approximation $\widetilde{\mathbf{x}}^H(n)\widetilde{\mathbf{x}}(n) = 2L\sigma_x^2$ for $L \gg 1$, knowing that $\mu(n)$ is deterministic in nature, substituting (4.72) into (4.71), using (4.70) to eliminate $\widetilde{\mathbf{h}}(n-1)$, and equating to (4.74), we find

$$\begin{aligned}E\left[|\epsilon(n)|^2\right] &= \left[1 - \mu(n)2L\sigma_x^2\right]^2 \sigma_e^2(n) \\ &= \sigma_v^2,\end{aligned} \tag{4.75}$$

where

$$\sigma_e^2(n) = E\left[|e(n)|^2\right] \tag{4.76}$$

is the variance of the error signal. Developing (4.75), we obtain the quadratic equation

$$\mu^2(n) - \frac{2}{2L\sigma_x^2}\mu(n) + \frac{1}{(2L\sigma_x^2)^2}\left[1 - \frac{\sigma_v^2}{\sigma_e^2(n)}\right] = 0, \tag{4.77}$$

for which the obvious solution is

$$\mu_{\text{VSS}}(n) = \frac{1}{\widetilde{\mathbf{x}}^H(n)\widetilde{\mathbf{x}}(n)}\left[1 - \frac{\sigma_v}{\sigma_e(n)}\right]$$
$$= \mu_{\text{NLMS}}(n)\alpha_{\text{VSS}}(n), \tag{4.78}$$

where $\alpha_{\text{VSS}}(n)$ $[0 \leq \alpha_{\text{VSS}}(n) \leq 1]$ is the normalized step-size parameter. Therefore, the VSS-NLMS algorithm is

$$\widetilde{\mathbf{h}}(n) = \widetilde{\mathbf{h}}(n-1) + \mu_{\text{VSS}}(n)\widetilde{\mathbf{x}}(n)e^*(n), \tag{4.79}$$

where $\mu_{\text{VSS}}(n)$ is defined in (4.78).

We see from (4.78) that before the algorithm converges, $\sigma_e(n)$ is large as compared to σ_v, thus $\mu_{\text{VSS}}(n) \approx \mu_{\text{NLMS}}(n)$. On the other hand, when the algorithm starts to converge, $\sigma_e(n) \approx \sigma_v$ and $\mu_{\text{VSS}}(n) \approx 0$. This is exactly what we desire to have both good convergence and low misadjustment. So in principle, the excess MSE and misadjustment of the VSS-NLMS should be

$$J_{\text{ex}}(\infty) \approx 0, \tag{4.80}$$
$$\mathcal{M} \approx 0. \tag{4.81}$$

In practice, (4.78) should be changed to

$$\mu_{\text{VSS}}(n) = \begin{cases} \dfrac{1}{\widetilde{\mathbf{x}}^H(n)\widetilde{\mathbf{x}}(n) + \delta}\left[1 - \dfrac{\sigma_v}{\widehat{\sigma}_e(n)}\right] & \text{if } \widehat{\sigma}_e(n) \geq \sigma_v \\ 0 & \text{otherwise} \end{cases}, \tag{4.82}$$

where the regularization parameter, δ, should be the same as for the NLMS,

$$\widehat{\sigma}_{\text{e}}^2(n) = \lambda_{\text{w}}\widehat{\sigma}_{\text{e}}^2(n-1) + (1 - \lambda_{\text{w}})\,|e(n)|^2 \tag{4.83}$$

is the estimation of the variance of the error signal, and $\lambda_{\text{w}} = 1 - \frac{1}{2C_{\text{w}}L}$ is an exponential window with $C_{\text{w}} \geq 3$.

As we can notice, this approach was derived with almost no assumptions as compared to all other algorithms belonging to the same family.

4.7 Improved Proportionate NLMS (IPNLMS) Algorithm

The sparseness character of the echo paths inspired the idea to "proportionate" the algorithm behavior, i.e., to update each coefficient of the filter independently of the others, by adjusting the adaptation step size in proportion to the magnitude of the estimated filter coefficient. In this manner, the adaptation gain is "proportionately" redistributed among all the coefficients, emphasizing the large ones in order to speed up their convergence, and consequently to increase the overall convergence rate. Even if the idea of exploiting the sparseness character of the systems appeared in the nineties, e.g., [17], [18], [19], the proportionate NLMS (PNLMS) algorithm [20] proposed by Duttweiler a decade ago, was one of the first "true" proportionate-type algorithms and maybe the most referenced one.

In this section, we show how to derive proportionate-type algorithms with basis pursuit (BP) [21].

In the BP approach, we try to solve an underdetermined linear system of equations, $\mathbf{b} = \mathbf{A}\mathbf{y}$, with the following optimization:

$$\min_{\mathbf{y}} \|\mathbf{y}\|_1 \quad \text{subject to} \quad \mathbf{b} = \mathbf{A}\mathbf{y}. \tag{4.84}$$

Under some conditions and if $\mathbf{y}$ is sufficiently sparse, it is possible to find the desired solution [21]. As a result, the minimum ℓ_1-norm solution is much more appropriate than the minimum ℓ_2-norm solution when the impulse response is sparse.

Now, going back to Section 4.4 in our interpretation of the NLMS algorithm and looking at (4.55), we see that we also have an underdetermined linear system to solve within the NLMS but the good thing about using an adaptive filter is that we do not need to solve the problem (i.e., converge to the optimal filter) in one iteration but, rather, in a finite number of iterations. Therefore, if the impulse response is sparse, it makes more sense to use BP in (4.51) instead of the minimum ℓ_2-norm solution, hoping that with BP less iterations will be required to converge to the same solution.

We can now derive a proportionate-type NLMS algorithm following the steps of our interpretation of the NLMS [5].

First, let us solve the optimization problem

$$\min_{\overleftarrow{\mathbf{h}}(n)} \left\|\overleftarrow{\mathbf{h}}(n)\right\|_1 \quad \text{subject to} \quad d(n) = \overleftarrow{\mathbf{h}}^H(n)\widetilde{\mathbf{x}}(n). \tag{4.85}$$

Using a Lagrange multiplier, we easily deduce that

$$\overleftarrow{\mathbf{h}}(n) = \frac{\overleftarrow{\mathbf{G}}(n)\widetilde{\mathbf{x}}(n)d^*(n)}{\widetilde{\mathbf{x}}^H(n)\overleftarrow{\mathbf{G}}(n)\widetilde{\mathbf{x}}(n)}, \tag{4.86}$$

where

$$\overleftarrow{\mathbf{G}}(n) = \text{diag}\left[\left|\overleftarrow{h}_0(n)\right|, \left|\overleftarrow{h}_1(n)\right|, \ldots, \left|\overleftarrow{h}_{2L-1}(n)\right|\right] \tag{4.87}$$

is a $2L \times 2L$ diagonal matrix and $\overleftarrow{h}_l(n)$, $l = 0, 1, \ldots, 2L-1$, are the components of $\overleftarrow{\mathbf{h}}(n)$.

Obviously, (4.86) is hard to solve. But since all elements of the diagonal matrix $\overleftarrow{\mathbf{G}}(n)$ are positive real numbers, approximating them by other positive components of the same nature is very reasonable and will certainly not affect the convergence of the adaptive algorithm to the optimal filter.

Since the coefficients, $\widetilde{h}_l(n-1),\ l=0,1,\ldots,2L-1$, of the adaptive filter of the previous time, $n-1$, are the best estimates available at time n, the most obvious and reliable approximation for $\overleftarrow{\mathbf{G}}(n)$ is

$$\begin{aligned}\overleftarrow{\mathbf{G}}(n) &\approx \mathbf{G}(n-1)\\ &= \mathrm{diag}\left[\left|\widetilde{h}_0(n-1)\right|, \left|\widetilde{h}_1(n-1)\right|, \ldots, \left|\widetilde{h}_{2L-1}(n-1)\right|\right]. \end{aligned} \tag{4.88}$$

Therefore, a good approximation of (4.86) in the context of adaptive filtering is

$$\overleftarrow{\mathbf{h}}(n) = \frac{\mathbf{G}(n-1)\widetilde{\mathbf{x}}(n)d^*(n)}{\widetilde{\mathbf{x}}^H(n)\mathbf{G}(n-1)\widetilde{\mathbf{x}}(n)}. \tag{4.89}$$

Second, we form a projection matrix for which $\widetilde{\mathbf{h}}(n)$ spans its nullspace. We propose

$$\mathbf{P}(n) = \mathbf{I}_{2L} - \frac{\mathbf{G}(n-1)\widetilde{\mathbf{x}}(n)\widetilde{\mathbf{x}}^H(n)}{\widetilde{\mathbf{x}}^H(n)\mathbf{G}(n-1)\widetilde{\mathbf{x}}(n)}. \tag{4.90}$$

It is easy to check that, indeed, $\mathbf{P}(n)$ is a (non-orthogonal) projection matrix since $\mathbf{P}^2(n) = \mathbf{P}(n)$ [but $\mathbf{P}^H(n) \neq \mathbf{P}(n)$].

Finally, the proportionate-type NLMS algorithm is

$$\widetilde{\mathbf{h}}(n) = \mathbf{P}(n)\widetilde{\mathbf{h}}(n-1) + \overleftarrow{\mathbf{h}}(n), \tag{4.91}$$

which can be rewritten as

$$\widetilde{\mathbf{h}}(n) = \widetilde{\mathbf{h}}(n-1) + \frac{\mathbf{G}(n-1)\widetilde{\mathbf{x}}(n)e^*(n)}{\widetilde{\mathbf{x}}^H(n)\mathbf{G}(n-1)\widetilde{\mathbf{x}}(n)}. \tag{4.92}$$

Making this algorithm more practical to avoid problems such as stalling of the coefficients, will lead to the well-known PNLMS and improved PNLMS (IPNLMS) algorithms [20], [22].

To make (4.92) more practical, we propose to write it as an IPNLMS form [22], i.e.,

$$\widetilde{\mathbf{h}}(n) = \widetilde{\mathbf{h}}(n-1) + \frac{\alpha\mathbf{G}(n-1)\widetilde{\mathbf{x}}(n)e^*(n)}{\widetilde{\mathbf{x}}^H(n)\mathbf{G}(n-1)\widetilde{\mathbf{x}}(n)+\delta}, \tag{4.93}$$

where α $(0 < \alpha < 2)$ is the normalized step-size parameter,

$$\mathbf{G}(n-1) = \mathrm{diag}\left[g_0(n-1), g_1(n-1), \ldots, g_{2L-1}(n-1)\right], \tag{4.94}$$

$$g_l(n-1) = \frac{1-\kappa}{4L} + (1+\kappa)\frac{\left|\widetilde{h}_l(n-1)\right|}{2\sum_{i=0}^{2L-1}\left|\widetilde{h}_i(n-1)\right|},\ 0 \leq l \leq 2L-1, \quad (4.95)$$

κ $(-1 \leq \kappa < 1)$ is a parameter that controls the amount of proportionality in the IPNLMS, and δ is the regularization constant.

4.8 Regularization of the IPNLMS Algorithm

The update equation of the IPNLMS can be rewritten as

$$\widetilde{\mathbf{h}}(n) = \mathbf{P}'(n)\widetilde{\mathbf{h}}(n-1) + \alpha\overleftarrow{\mathbf{h}}'(n), \quad (4.96)$$

where now

$$\mathbf{P}'(n) = \mathbf{I}_{2L} - \alpha\frac{\mathbf{G}(n-1)\widetilde{\mathbf{x}}(n)\widetilde{\mathbf{x}}^H(n)}{\widetilde{\mathbf{x}}^H(n)\mathbf{G}(n-1)\widetilde{\mathbf{x}}(n)+\delta} \quad (4.97)$$

and

$$\overleftarrow{\mathbf{h}}'(n) = \frac{\mathbf{G}(n-1)\widetilde{\mathbf{x}}(n)d^*(n)}{\widetilde{\mathbf{x}}^H(n)\mathbf{G}(n-1)\widetilde{\mathbf{x}}(n)+\delta}, \quad (4.98)$$

which depends on the new observation $d(n)$. Note that $\mathbf{P}'(n)$ does not depend on the noise signal.

It can be shown that $\overleftarrow{\mathbf{h}}'(n)$ is a good approximation of the optimization problem:

$$\min_{\overleftarrow{\mathbf{h}}'(n)} \left[\left|d(n) - \overleftarrow{\mathbf{h}}'^H(n)\widetilde{\mathbf{x}}(n)\right|^2 + \delta\left\|\overleftarrow{\mathbf{h}}'(n)\right\|_1\right]. \quad (4.99)$$

The previous optimization is the regularized version of the minimum ℓ_1-norm solution of the linear system of one equation $d(n) = \overleftarrow{\mathbf{h}}'^H(n)\widetilde{\mathbf{x}}(n)$. Therefore, we can use the condition (4.61) to derive δ.

For $L \gg 1$ and a stationary signal $x(n)$, we have

$$\begin{aligned}\widetilde{\mathbf{x}}^H(n)\mathbf{G}(n-1)\widetilde{\mathbf{x}}(n) &= \frac{1-\kappa}{4L}\widetilde{\mathbf{x}}^H(n)\widetilde{\mathbf{x}}(n) \\ &\quad + \frac{1+\kappa}{2\left\|\widetilde{\mathbf{h}}(n-1)\right\|_1}\sum_{l=0}^{2L-1}\left|\widetilde{x}_l(n)\right|^2\left|\widetilde{h}_l(n-1)\right| \\ &\approx \frac{1-\kappa}{2}\sigma_x^2 + \frac{1+\kappa}{2}\sigma_x^2 \\ &\approx \sigma_x^2, \end{aligned} \quad (4.100)$$

where $\widetilde{x}_l(n),\ l = 0, 1, \ldots, 2L-1$, are the components of the vector $\widetilde{\mathbf{x}}(n)$.

Using the condition (4.61) and (4.100), we easily derive the quadratic equation

$$\delta^2 - 2\frac{\sigma_x^2}{\text{SENR}}\delta - \frac{\left(\sigma_x^2\right)^2}{\text{SENR}} = 0, \tag{4.101}$$

from which the desired solution is

$$\begin{aligned}\delta &= \frac{1+\sqrt{1+\text{SENR}}}{\text{SENR}}\sigma_x^2 \\ &= \beta_{\text{IPNLMS}}\sigma_x^2,\end{aligned} \tag{4.102}$$

where

$$\beta_{\text{IPNLMS}} = \frac{1+\sqrt{1+\text{SENR}}}{\text{SENR}} \tag{4.103}$$

is the normalized regularization parameter of the IPNLMS.

It is interesting to observe that the regularization does not depend on the parameter κ. In fact, the regularization of the IPNLMS is equivalent to the regularization of the NLMS up to the scaling factor $2L$, which is due to the definition of $g_l(n-1)$.

4.9 VSS-IPNLMS Algorithm

In order to approach the goal of finding a VSS-IPNLMS algorithm [23] suitable for any kind of impulse response, we will apply the idea of VSS-NLMS developed in Section 4.6 to the IPNLMS. Let us write the update equation of the IPNLMS as follows:

$$\widetilde{\mathbf{h}}(n) = \widetilde{\mathbf{h}}(n-1) + \mu_{\text{VSS}}(n)\mathbf{G}(n-1)\widetilde{\mathbf{x}}(n)e^*(n), \tag{4.104}$$

where $\mathbf{G}(n-1)$ is a diagonal matrix with its elements defined in (4.95). The issue is to find an expression for the step-size parameter, $\mu_{\text{VSS}}(n)$, according to the constraint

$$\begin{aligned}E\left[|\epsilon(n)|^2\right] &= E\left[\left|d(n) - \widetilde{\mathbf{h}}^H(n)\widetilde{\mathbf{x}}(n)\right|^2\right] \\ &= \sigma_v^2.\end{aligned} \tag{4.105}$$

Following the same procedure as for the derivation of the VSS-NLMS algorithm, it will result that

$$\mu_{\text{VSS}}(n) = \begin{cases} \dfrac{1}{\widetilde{\mathbf{x}}^H(n)\mathbf{G}(n-1)\widetilde{\mathbf{x}}(n)+\delta}\left[1-\dfrac{\sigma_v}{\widehat{\sigma}_e(n)}\right] & \text{if } \widehat{\sigma}_e(n) \geq \sigma_v \\ 0 & \text{otherwise} \end{cases}, \tag{4.106}$$

where the regularization parameter, δ, should be the same as for the IPNLMS and $\widehat{\sigma}_e^2(n)$ is estimated as in (4.83).

4.10 Extended NLMS (ENLMS) Algorithm

The extended NLMS (ENLMS) algorithm was first proposed in [24]. In the WL context, we define the update equation of the ENLMS as

$$\widetilde{\mathbf{h}}(n) = \widetilde{\mathbf{h}}(n-1) + \frac{1}{2}\begin{bmatrix} \mathbf{x}^H(n)\mathbf{x}(n)\mathbf{I}_L & \mathbf{x}^T(n)\mathbf{x}(n)\mathbf{I}_L \\ \mathbf{x}^H(n)\mathbf{x}^*(n)\mathbf{I}_L & \mathbf{x}^H(n)\mathbf{x}(n)\mathbf{I}_L \end{bmatrix}^{-1} \widetilde{\mathbf{x}}(n)e^*(n), \tag{4.107}$$

where $\mathbf{I}_L$ is the $L \times L$ identity matrix. It can be checked that this update equation cancels the a posteriori error signal, i.e.,

$$\epsilon(n) = d(n) - \widetilde{\mathbf{h}}^H(n)\widetilde{\mathbf{x}}(n) = 0. \tag{4.108}$$

Another way to express (4.107) is

$$\widetilde{\mathbf{h}}(n) = \widetilde{\mathbf{h}}(n-1) + \frac{1}{\widetilde{\mathbf{x}}^H(n)\widetilde{\mathbf{x}}(n)}\begin{bmatrix} \mathbf{I}_L & \widehat{\gamma}_x(n)\mathbf{I}_L \\ \widehat{\gamma}_x^*(n)\mathbf{I}_L & \mathbf{I}_L \end{bmatrix}^{-1} \widetilde{\mathbf{x}}(n)e^*(n), \tag{4.109}$$

where

$$\begin{aligned} \widehat{\gamma}_x(n) &= \frac{\mathbf{x}^T(n)\mathbf{x}(n)}{\mathbf{x}^H(n)\mathbf{x}(n)} \\ &\approx \frac{E\left[x^2(n)\right]}{\sigma_x^2} \end{aligned} \tag{4.110}$$

is an estimate of the circularity quotient. We observe that if $x(n)$ is a second-order CCRV [i.e., $\widehat{\gamma}_x(n) = 0$] then the ENLMS simplifies to the NLMS.

We can also update the two sub-filters of $\widetilde{\mathbf{h}}(n) = \left[\mathbf{h}(n)\ \mathbf{h}'(n)\right]^T$ separately as

$$\mathbf{h}(n) = \mathbf{h}(n-1) + \alpha_x\left\{\left[\mathbf{x}^H(n)\mathbf{x}(n)\right]\mathbf{x}(n) - \left[\mathbf{x}^T(n)\mathbf{x}(n)\right]\mathbf{x}^*(n)\right\}e^*(n), \tag{4.111}$$

$$\mathbf{h}'(n) = \mathbf{h}'(n-1) + \alpha_x\left\{\left[\mathbf{x}^H(n)\mathbf{x}(n)\right]\mathbf{x}^*(n) - \left[\mathbf{x}^H(n)\mathbf{x}^*(n)\right]\mathbf{x}(n)\right\}e^*(n),$$

$$(4.112)$$

where

$$\alpha_x = \frac{0.5}{\left[\mathbf{x}^H(n)\mathbf{x}(n)\right]^2 - \left[\mathbf{x}^T(n)\mathbf{x}(n)\right]\left[\mathbf{x}^H(n)\mathbf{x}^*(n)\right]}. \tag{4.113}$$

To make the ENLMS more practical, it is preferable to rewrite it as follows:

$$\widetilde{\mathbf{h}}(n) = \widetilde{\mathbf{h}}(n-1) + \frac{\alpha}{\widetilde{\mathbf{x}}^H(n)\widetilde{\mathbf{x}}(n) + \delta} \begin{bmatrix} \mathbf{I}_L & \widehat{\gamma}_x(n)\mathbf{I}_L \\ \widehat{\gamma}_x^*(n)\mathbf{I}_L & \mathbf{I}_L \end{bmatrix}^{-1} \widetilde{\mathbf{x}}(n)e^*(n), \tag{4.114}$$

where α $(0 < \alpha < 2)$ is the normalized step-size parameter, δ is the regularization constant, which should have the same value as for the NLMS, and now

$$\widehat{\gamma}_x(n) = \frac{\mathbf{x}^T(n)\mathbf{x}(n)}{\mathbf{x}^H(n)\mathbf{x}(n) + \delta}. \tag{4.115}$$

Finally, to end this part, we give the VSS version of the ENLMS:

$$\widetilde{\mathbf{h}}(n) = \widetilde{\mathbf{h}}(n-1) + \mu_{\mathrm{VSS}}(n) \begin{bmatrix} \mathbf{I}_L & \widehat{\gamma}_x(n)\mathbf{I}_L \\ \widehat{\gamma}_x^*(n)\mathbf{I}_L & \mathbf{I}_L \end{bmatrix}^{-1} \widetilde{\mathbf{x}}(n)e^*(n), \tag{4.116}$$

where $\mu_{\mathrm{VSS}}(n)$ is defined in (4.82).

References

1. B. Widrow and M. E. Hoff, Jr., "Adaptive switching circuits," *IRE WESCON Conv. Rec.*, Pt. 4, 1960, pp. 96–104.
2. S. Haykin, *Adaptive Filter Theory.* Fourth Edition, Upper Saddle River, NJ: Prentice-Hall, 2002.
3. B. Widrow, J. M. McCool, M. G. Larimore, and C. R. Johnson, Jr., "Stationary and nonstationary learning characteristics of the LMS adaptive filter," *Proc. of the IEEE*, vol. 64, pp. 1151–1162, Aug. 1976.
4. J.-I. Nagumo and A. Noda, "A learning method for system identification," *IEEE Trans. Autom. Control*, vol. AC-12, pp. 282–287, June 1967.
5. J. Benesty, C. Paleologu, and S. Ciochină, "Proportionate adaptive filters from a basis pursuit perspective," *IEEE Signal Process. Lett.*, vol. 17, pp. 985–988, Dec. 2010.
6. J. Benesty, C. Paleologu, and S. Ciochină, "On regularization in adaptive filtering," *IEEE Trans. Audio, Speech, Language Process.*, vol. 19, pp. 1734–1742, Aug. 2011.
7. R. W. Harris, D. M. Chabries, and F. A. Bishop, "A variable step (VS) adaptive filter algorithm," *IEEE Trans. Acoust., Speech, Signal Process.*, vol. ASSP-34, pp. 309–316, Apr. 1986.
8. R. H. Kwong and E. W. Johnston, "A variable step size LMS algorithm," *IEEE Trans. Signal Process.*, vol. 40, pp. 1633–1642, July 1992.

9. V. J. Mathews and Z. Xie, "A stochastic gradient adaptive filter with gradient adaptive step size," *IEEE Trans. Signal Process.*, vol. 41, pp. 2075–2087, June 1993.
10. J. B. Evans, P. Xue, and B. Liu, "Analysis and implementation of variable step size adaptive algorithms," *IEEE Trans. Signal Process.*, vol. 41, pp. 2517–2535, Aug. 1993.
11. T. Aboulnasr and K. Mayyas, "A robust variable step-size LMS-type algorithm: analysis and simulations," *IEEE Trans. Signal Process.*, vol. 45, pp. 631–639, Mar. 1997.
12. D. I. Pazaitis and A. G. Constantinides, "A novel kurtosis driven variable step-size adaptive algorithm," *IEEE Trans. Signal Process.*, vol. 47, pp. 864–872, Mar. 1999.
13. A. Mader, H. Puder, and G. U. Schmidt, "Step-size control for acoustic echo cancellation filters – An overview," *Signal Process.*, vol. 80, pp. 1697–1719, Sept. 2000.
14. H.-C. Shin, A. H. Sayed, and W.-J. Song, "Variable step-size NLMS and affine projection algorithms," *IEEE Signal Process. Lett.*, vol. 11, pp. 132–135, Feb. 2004.
15. J. Benesty, H. Rey, L. Rey Vega, and S. Tressens, "A non-parametric VSS-NLMS algorithm," *IEEE Signal Process. Lett.*, vol. 13, pp. 581–584, Oct. 2006.
16. D. R. Morgan and S. G. Kratzer, "On a class of computationally efficient, rapidly converging, generalized NLMS algorithms," *IEEE Signal Process. Lett.*, vol. 3, pp. 245–247, Aug. 1996.
17. S. Makino, Y. Kaneda, and N. Koizumi, "Exponentially weighted step-size NLMS adaptive filter based on the statistics of a room impulse response," *IEEE Trans. Speech, Audio Process.*, vol. 1, pp. 101–108, Jan. 1993.
18. A. Sugiyama, H. Sato, A. Hirano, and S. Ikeda, "A fast convergence algorithm for adaptive FIR filters under computational constraint for adaptive tap-position control," *IEEE Trans. Circuits Syst. II*, vol. 43, pp. 629–636, Sept. 1996.
19. J. Homer, I. Mareels, R. R. Bitmead, B. Wahlberg, and A. Gustafsson, "LMS estimation via structural detection," *IEEE Trans. Signal Process.*, vol. 46, pp. 2651–2663, Oct. 1998.
20. D. L. Duttweiler, "Proportionate normalized least-mean-squares adaptation in echo cancelers," *IEEE Trans. Speech, Audio Process.*, vol. 8, pp. 508–518, Sept. 2000.
21. S. Chen, D. Donoho, and M. Saunders, "Atomic decomposition by basis pursuit," *SIAM J. Sci. Comput.*, vol. 20, no. 1, pp. 33–61, 1998.
22. J. Benesty and S. L. Gay, "An improved PNLMS algorithm," in *Proc. IEEE ICASSP*, 2002, pp. 1881–1884.
23. C. Paleologu, J. Benesty, and S. Ciochină, "A variable step-size proportionate NLMS algorithm for echo cancellation," *Revue Roumaine des Sciences Techniques – Serie Electrotechnique et Energetique*, vol. 53, no. 3, pp. 309–317, 2008.
24. J. Benesty, F. Amand, A. Gilloire, and Y. Grenier, "Adaptive filtering algorithms for stereophonic acoustic echo cancellation," in *Proc. IEEE ICASSP*, 1995, pp. 3099–3102.

Chapter 5
A Class of Affine Projection Algorithms

Affine projection algorithms (APAs) are very good candidates for echo cancellation. The two main reasons for that are: they may converge and track much faster than the NLMS algorithm and they can be efficient from an arithmetic complexity viewpoint. In this chapter, we derive some useful APAs for SAEC with the WL model.

5.1 Affine Projection Algorithm (APA)

In this section, we briefly derive the affine projection algorithm (APA) that was originally proposed in [1]. We develop it in the WL context. As a consequence, the resulting algorithm is equivalent to the one proposed in [2]. The APA can be seen as a generalization of the NLMS, where the projection order is increased from 1 to P. The clear advantage of the APA over the NLMS is that it may converge must faster thanks to the fact it considers in its update equation both past and current input vectors.

Let us write the $2L \times P$ input matrix

$$\widetilde{\mathbf{X}}(n) = \left[\widetilde{\mathbf{x}}(n) \; \widetilde{\mathbf{x}}(n-1) \cdots \widetilde{\mathbf{x}}(n-P+1) \right].$$

We define the $P \times 1$ a priori error vector as

$$\mathbf{e}(n) = \mathbf{d}(n) - \widetilde{\mathbf{X}}^T(n)\widetilde{\mathbf{h}}^*(n-1), \tag{5.1}$$

where

$$\mathbf{d}(n) = \left[d(n) \; d(n-1) \cdots d(n-P+1) \right]^T$$

is a vector containing the most recent P samples of the output signal. We see that the vector $\mathbf{e}(n)$ contains the errors for the current and past $P-1$ input signal vectors, which are used to update the filter at time n. In the same way,

J. Benesty et al.: A Perspective on Stereophonic Acoustic Echo Cancellation, STSP 4, pp. 49–62.
springerlink.com

we define the $P \times 1$ a posteriori error vector as

$$\boldsymbol{\epsilon}(n) = \mathbf{d}(n) - \widetilde{\mathbf{X}}^T(n)\widetilde{\mathbf{h}}^*(n). \tag{5.2}$$

The APA is obtained as follows:

$$\min_{\widetilde{\mathbf{h}}(n)} \left\| \widetilde{\mathbf{h}}(n) - \widetilde{\mathbf{h}}(n-1) \right\|_2^2 \quad \text{subject to} \quad \boldsymbol{\epsilon}(n) = \mathbf{0}. \tag{5.3}$$

Using Lagrange multipliers, we easily find the update equation of the algorithm, which is

$$\widetilde{\mathbf{h}}(n) = \widetilde{\mathbf{h}}(n-1) + \widetilde{\mathbf{X}}(n) \left[\widetilde{\mathbf{X}}^H(n)\widetilde{\mathbf{X}}(n) \right]^{-1} \mathbf{e}^*(n). \tag{5.4}$$

It can be checked that, indeed, (5.4) cancels the a posteriori error vector. Also, by taking $P = 1$, we find the NLMS algorithm derived in the previous chapter.

To make the APA more practical, we need to change it a bit by including the two classical parameters: the normalized step size, α ($0 < \alpha < 2$), and the regularization, $\delta \geq 0$. Therefore, the following three equations summarize this algorithm:

$$\widehat{\mathbf{y}}(n) = \widetilde{\mathbf{X}}^T(n)\widetilde{\mathbf{h}}^*(n-1), \text{ filter output vector,} \tag{5.5}$$

$$\mathbf{e}(n) = \mathbf{d}(n) - \widehat{\mathbf{y}}(n), \text{ error vector,} \tag{5.6}$$

$$\widetilde{\mathbf{h}}(n) = \widetilde{\mathbf{h}}(n-1) + \alpha\widetilde{\mathbf{X}}(n) \left[\widetilde{\mathbf{X}}^H(n)\widetilde{\mathbf{X}}(n) + \delta\mathbf{I}_P \right]^{-1} \mathbf{e}^*(n), \text{ adaptation,} \tag{5.7}$$

where $\mathbf{I}_P$ is the $P \times P$ identity matrix.

5.2 Interpretation of the APA

In this section, we assume that $\alpha = 1$ and $\delta = 0$. In this case, the update equation of the APA can be rewritten as [3]

$$\widetilde{\mathbf{h}}(n) = \mathbf{P}(n)\widetilde{\mathbf{h}}(n-1) + \overleftarrow{\mathbf{h}}(n), \tag{5.8}$$

where

$$\mathbf{P}(n) = \mathbf{I}_{2L} - \widetilde{\mathbf{X}}(n) \left[\widetilde{\mathbf{X}}^H(n)\widetilde{\mathbf{X}}(n) \right]^{-1} \widetilde{\mathbf{X}}^H(n) \tag{5.9}$$

is an orthogonal projection matrix whose rank is equal to $2L - P$ and

$$\overleftarrow{\mathbf{h}}(n) = \widetilde{\mathbf{X}}(n) \left[\widetilde{\mathbf{X}}^H(n)\widetilde{\mathbf{X}}(n)\right]^{-1} \mathbf{d}^*(n) \tag{5.10}$$

is the correction component component of the APA. We observe that $\widetilde{\mathbf{h}}(n)$ is the sum of two vectors that are orthogonal, i.e.,

$$\left[\mathbf{P}(n)\widetilde{\mathbf{h}}(n-1)\right]^H \overleftarrow{\mathbf{h}}(n) = 0. \tag{5.11}$$

Therefore, $\overleftarrow{\mathbf{h}}(n)$ belongs to the nullspace of $\mathbf{P}(n)$, i.e.,

$$\mathbf{P}(n)\overleftarrow{\mathbf{h}}(n) = \mathbf{0}. \tag{5.12}$$

Furthermore $\widetilde{\mathbf{h}}(n-1)$, in the space formed from the complementary (range and null) subspaces of $\mathbf{P}(n)$, can be decomposed uniquely as

$$\widetilde{\mathbf{h}}(n-1) = \widetilde{\mathbf{h}}_{\parallel}(n-1) + \widetilde{\mathbf{h}}_{\perp}(n-1), \tag{5.13}$$

where

$$\mathbf{P}(n)\widetilde{\mathbf{h}}(n-1) = \widetilde{\mathbf{h}}_{\parallel}(n-1), \tag{5.14}$$

$$\mathbf{P}(n)\widetilde{\mathbf{h}}_{\perp}(n-1) = \mathbf{0}. \tag{5.15}$$

As a result, (5.8) simplifies to

$$\widetilde{\mathbf{h}}(n) = \widetilde{\mathbf{h}}_{\parallel}(n-1) + \overleftarrow{\mathbf{h}}(n), \tag{5.16}$$

since the component $\widetilde{\mathbf{h}}_{\perp}(n-1)$ does not affect the update equation. Clearly, we have $\widetilde{\mathbf{h}}_{\parallel}^H(n-1)\overleftarrow{\mathbf{h}}(n) = 0$ and $\widetilde{\mathbf{X}}^H(n)\widetilde{\mathbf{h}}(n) = \widetilde{\mathbf{X}}^H(n)\overleftarrow{\mathbf{h}}(n) = \mathbf{d}^*(n)$. According to (5.13) and (5.14),

$$\widetilde{\mathbf{h}}_{\perp}(n-1) = \left[\mathbf{I}_{2L} - \mathbf{P}(n)\right]\widetilde{\mathbf{h}}(n-1). \tag{5.17}$$

Then, using (5.9) in (5.17), we get

$$\widetilde{\mathbf{h}}_{\perp}(n-1) = \widetilde{\mathbf{X}}(n) \left[\widetilde{\mathbf{X}}^H(n)\widetilde{\mathbf{X}}(n)\right]^{-1} \widehat{\mathbf{y}}^*(n), \tag{5.18}$$

which is interesting to compare to $\overleftarrow{\mathbf{h}}(n)$. Both vectors are obviously in the nullspace of $\mathbf{P}(n)$. Another way, then, to write the error signal is

$$\mathbf{e}(n) = \widetilde{\mathbf{X}}^T(n) \left[\overleftarrow{\mathbf{h}}^*(n) - \widetilde{\mathbf{h}}_{\perp}^*(n-1)\right], \tag{5.19}$$

which is not affected by $\widetilde{\mathbf{h}}_{\parallel}(n-1)$.

The most interesting part in the decomposition of $\widetilde{\mathbf{h}}(n)$ in (5.16) is that $\overleftarrow{\mathbf{h}}(n)$ is, in fact, the minimum ℓ_2-norm solution of the linear system of P

equations $\mathbf{d}(n) = \widetilde{\mathbf{X}}^T(n)\overleftarrow{\mathbf{h}}^*(n)$. Indeed, by

$$\min_{\overleftarrow{\mathbf{h}}(n)} \left\| \overleftarrow{\mathbf{h}}(n) \right\|_2^2 \quad \text{subject to} \quad \mathbf{d}(n) = \widetilde{\mathbf{X}}^T(n)\overleftarrow{\mathbf{h}}^*(n), \tag{5.20}$$

we find (5.10). And, of course, $\widetilde{\mathbf{h}}(n)$ is also a solution of $\mathbf{d}(n) = \widetilde{\mathbf{X}}^T(n)\overleftarrow{\mathbf{h}}^*(n)$. Therefore, $\widetilde{\mathbf{h}}_{\|}(n-1)$ can be seen as a good initialization of the adaptive filter since the minimum ℓ_2-norm solution is not the optimal solution. Obviously, for $P = 1$, we get the NLMS algorithm.

The interpretation [3] of the APA is the following. First, we find the minimum ℓ_2-norm solution of $\mathbf{d}(n) = \widetilde{\mathbf{X}}^T(n)\overleftarrow{\mathbf{h}}^*(n)$, which is denoted by $\overleftarrow{\mathbf{h}}(n)$. This step is the most important one since it determines the adaptation with the most recent information. Second, we form an (orthogonal) projection matrix, $\mathbf{P}(n)$, of rank $2L - P$ for which $\overleftarrow{\mathbf{h}}(n)$ spans its nullspace. Third, from $\widetilde{\mathbf{h}}(n-1)$ we find the initialization vector, $\widetilde{\mathbf{h}}_{\|}(n-1)$, that is in the range of $\mathbf{P}(n)$. Finally, we add $\overleftarrow{\mathbf{h}}(n)$ and $\widetilde{\mathbf{h}}_{\|}(n-1)$ to find the new update equation resulting in the APA.

5.3 Regularization of the APA

It is extremely important to properly regularize the APA in order to expect good performance in different scenarios of noise and input signals.

The value of the regularization parameter δ depends on the level of the noise. Low SENRs require high values of the regularization parameter, while its importance becomes less apparent for high SENRs. This is a rule of thumb in practice, but the question is how large or small δ should be chosen as a function of the SENR?

In order to provide a possible answer to this question, let us first rewrite (5.7) as [4]

$$\widetilde{\mathbf{h}}(n) = \mathbf{P}'(n)\widetilde{\mathbf{h}}(n-1) + \alpha\overleftarrow{\mathbf{h}}'(n), \tag{5.21}$$

where now

$$\mathbf{P}'(n) = \mathbf{I}_{2L} - \alpha\widetilde{\mathbf{X}}(n)\left[\widetilde{\mathbf{X}}^H(n)\widetilde{\mathbf{X}}(n) + \delta\mathbf{I}_P\right]^{-1}\widetilde{\mathbf{X}}^H(n) \tag{5.22}$$

and

$$\overleftarrow{\mathbf{h}}'(n) = \widetilde{\mathbf{X}}(n)\left[\widetilde{\mathbf{X}}^H(n)\widetilde{\mathbf{X}}(n) + \delta\mathbf{I}_P\right]^{-1}\mathbf{d}^*(n). \tag{5.23}$$

It is important to interpret the significance of the terms in (5.21). First, the vector $\overleftarrow{\mathbf{h}}'(n)$ is the correction component of the APA, since it depends on the new observation $\mathbf{d}(n)$. On the other hand, the matrix $\mathbf{P}'(n)$ does not depend on the noise signal or the desired signal (it depends only on the input signal). Second, it can be noticed that $\overleftarrow{\mathbf{h}}'(n)$ is obtained by solving

$$\min_{\overleftarrow{\mathbf{h}}'(n)} \left[\|\overleftarrow{\mathbf{e}}(n)\|_2^2 + \delta \left\| \overleftarrow{\mathbf{h}}'(n) \right\|_2^2 \right], \tag{5.24}$$

where

$$\overleftarrow{\mathbf{e}}(n) = \mathbf{d}(n) - \widetilde{\mathbf{X}}^T(n) \overleftarrow{\mathbf{h}}'^*(n) \tag{5.25}$$

is the error vector between the desired signal vector and the estimated signal vector obtained from $\overleftarrow{\mathbf{h}}'(n)$. The optimization from (5.24) is the regularized version of the minimum ℓ_2-norm solution of the linear system of P equations $\mathbf{d}(n) = \widetilde{\mathbf{X}}^T(n)\overleftarrow{\mathbf{h}}'^*(n)$. Consequently, since the solution $\overleftarrow{\mathbf{h}}'(n)$ is not the optimal one, the other vector $\mathbf{P}'(n)\widetilde{\mathbf{h}}(n-1)$ in (5.21) can be seen as a good initialization of the adaptive filter.

We can rewrite the a posteriori error vector as

$$\begin{aligned} \boldsymbol{\epsilon}(n) &= \mathbf{d}(n) - \widetilde{\mathbf{X}}^T(n)\widetilde{\mathbf{h}}^*(n) \\ &= \widetilde{\mathbf{X}}^T(n) \left[\widetilde{\mathbf{h}}_t^* - \widetilde{\mathbf{h}}^*(n) \right] + \mathbf{v}(n), \end{aligned} \tag{5.26}$$

where

$$\mathbf{v}(n) = \begin{bmatrix} v(n) \; v(n-1) \; \cdots \; v(n-P+1) \end{bmatrix}^T$$

is a vector containing the most recent P samples of the system noise. It is known that the basic idea of the APA is to cancel P a posteriori errors [1], i.e., $\boldsymbol{\epsilon}(n) = \mathbf{0}$. Considering the unregularized version of the APA [i.e., neglecting the regularization parameter in (5.7)] and replacing (5.7) in the first line of (5.26) [taking (5.6) into account], it results that $\boldsymbol{\epsilon}(n) = (1-\alpha)\mathbf{e}(n)$. Consequently, assuming that $\mathbf{e}(n) \neq \mathbf{0}$, in order to cancel P a posteriori errors, the step size should be $\alpha = 1$. However, as it can be noticed from the second line of (5.26), this approach holds only in the absence of the noise, i.e., $\mathbf{v}(n) = 0$. Otherwise, cancelling the a posteriori error vector will introduce noise in the adaptive filter estimate. In order to overcome this issue, the recently proposed variable-regularized APA in [5] follows the condition

$$E\left[\|\boldsymbol{\epsilon}(n)\|_2^2 \right] = E\left[\|\mathbf{v}(n)\|_2^2 \right] \tag{5.27}$$

to find a variable regularization parameter.

On the other hand, as we can see from (5.21)–(5.23), the first term of the update (5.21), i.e., $\mathbf{P}'(n)\widetilde{\mathbf{h}}(n-1)$, does not depend on the noise signal. The influence of the noise can be found in the second term of (5.21), i.e., mainly in the estimator $\overleftarrow{\mathbf{h}}'(n)$ and, consequently, in the error $\overleftarrow{\mathbf{e}}(n)$ defined in (5.25). Therefore, in order to attenuate the effects of the noise in the estimator $\overleftarrow{\mathbf{h}}'(n)$, it is reasonable to find δ in such a way that [4]

$$E\left[\|\overleftarrow{\mathbf{e}}(n)\|_2^2\right] = E\left[\|\mathbf{v}(n)\|_2^2\right]. \tag{5.28}$$

In order to further develop the previous condition, we can use (5.23) in (5.25) to get

$$\overleftarrow{\mathbf{e}}(n) = \left\{\mathbf{I}_P - \widetilde{\mathbf{X}}^T(n)\widetilde{\mathbf{X}}^*(n)\left[\widetilde{\mathbf{X}}^T(n)\widetilde{\mathbf{X}}^*(n) + \delta\mathbf{I}_P\right]^{-1}\right\}\mathbf{d}(n). \tag{5.29}$$

Also, let us use the eigenvalue decomposition:

$$\widetilde{\mathbf{X}}^T(n)\widetilde{\mathbf{X}}^*(n) = \mathbf{Q}_P(n)\mathbf{\Lambda}_P(n)\mathbf{Q}_P^H(n), \tag{5.30}$$

where $\mathbf{Q}_P(n)$ is a unitary matrix containing the eigenvectors of $\widetilde{\mathbf{X}}^T(n)\widetilde{\mathbf{X}}^*(n)$ as columns and $\mathbf{\Lambda}_P(n)$ is a diagonal matrix containing the corresponding eigenvalues $\lambda_{P,p}(n)$, with $p = 0, 1, \ldots, P-1$. Consequently,

$$\left[\widetilde{\mathbf{X}}^T(n)\widetilde{\mathbf{X}}^*(n) + \delta\mathbf{I}_P\right]^{-1} = \mathbf{Q}_P(n)\left[\delta\mathbf{I}_P + \mathbf{\Lambda}_P(n)\right]^{-1}\mathbf{Q}_P^H(n). \tag{5.31}$$

Next, based on (5.29)–(5.31), we get

$$\|\overleftarrow{\mathbf{e}}(n)\|_2^2 = \mathbf{d}^H(n)\mathbf{Q}_P(n)\left\{\mathbf{I}_P - \mathbf{\Lambda}_P(n)\left[\delta\mathbf{I}_P + \mathbf{\Lambda}_P(n)\right]^{-1}\right\}^2\mathbf{Q}_P^H(n)\mathbf{d}(n). \tag{5.32}$$

It is very difficult to further process (5.32) without any supporting assumptions on the character of the input signal. In order to facilitate the analysis, let us assume that the input signal is white, so that the eigenvalues of the matrix $\widetilde{\mathbf{X}}^T(n)\widetilde{\mathbf{X}}^*(n)$ [see (5.30)] are

$$\lambda_{P,p}(n) \approx 2L\sigma_x^2, \text{ for } p = 0, 1, \ldots, P-1. \tag{5.33}$$

Therefore, (5.32) can be expressed as

$$\|\overleftarrow{\mathbf{e}}(n)\|_2^2 = \mathbf{d}^H(n)\mathbf{Q}_P(n)\mathbf{Q}_P^H(n)\mathbf{d}(n)\left(\frac{\delta}{\delta + 2L\sigma_x^2}\right)^2. \tag{5.34}$$

Consequently, taking the expectation on both sides of (5.34), the condition (5.28) becomes

$$E\left[\|\mathbf{d}(n)\|_2^2\right]\left(\frac{\delta}{\delta + 2L\sigma_x^2}\right)^2 = E\left[\|\mathbf{v}(n)\|_2^2\right]. \tag{5.35}$$

Since the echo signal and the system noise are uncorrelated,

$$E\left[\|\mathbf{d}(n)\|_2^2\right] = E\left[\|\mathbf{y}(n)\|_2^2\right] + E\left[\|\mathbf{v}(n)\|_2^2\right], \tag{5.36}$$

where

$$\mathbf{y}(n) = \left[y(n)\ y(n-1)\ \cdots\ y(n-P+1)\right]^T$$

contains the most recent P samples of the echo signal. Finally, using the relations $E\left[\|\mathbf{y}(n)\|_2^2\right] = P\sigma_y^2$ and $E\left[\|\mathbf{v}(n)\|_2^2\right] = P\sigma_v^2$, and based on the definition of the SENR, the condition (5.35) becomes

$$\left(\frac{\delta}{\delta + 2L\sigma_x^2}\right)^2 = \frac{1}{1 + \mathrm{SENR}}, \tag{5.37}$$

which results in

$$\mathrm{SENR} \cdot \delta^2 - 2\left(2L\sigma_x^2\right)\delta - \left(2L\sigma_x^2\right)^2 = 0. \tag{5.38}$$

The obvious solution of the quadratic equation (5.38) is

$$\begin{aligned} \delta &= \frac{2L\left(1 + \sqrt{1 + \mathrm{SENR}}\right)}{\mathrm{SENR}}\sigma_x^2 \\ &= \beta_{\mathrm{APA}}\sigma_x^2, \end{aligned} \tag{5.39}$$

where

$$\beta_{\mathrm{APA}} = \frac{2L\left(1 + \sqrt{1 + \mathrm{SENR}}\right)}{\mathrm{SENR}} \tag{5.40}$$

is the normalized (with respect to the variance of the input signal) regularization parameter of the APA.

It can be noticed that the regularization parameter of the APA does not depend on P and is identical to the regularization parameter of the NLMS algorithm when we assume that the input signal is white. In the general case, the expression of the regularization is much more complicated. However, in practice, the regularization parameter needs not to be accurate; an approximate value gives, usually, good performance.

5.4 Variable Step-Size APA (VSS-APA)

A variable step-size version (VSS) of the APA is always interesting in echo cancellation. For convenience of the derivation, we start by neglecting the regularization matrix $\delta\mathbf{I}_P$ in (5.7), so that the update of the APA becomes

$$\widetilde{\mathbf{h}}(n) = \widetilde{\mathbf{h}}(n-1) + \alpha\widetilde{\mathbf{X}}(n)\left[\widetilde{\mathbf{X}}^H(n)\widetilde{\mathbf{X}}(n)\right]^{-1}\mathbf{e}^*(n). \tag{5.41}$$

Next, let us rewrite the previous update as

$$\widetilde{\mathbf{h}}(n) = \widetilde{\mathbf{h}}(n-1) + \widetilde{\mathbf{X}}(n)\left[\widetilde{\mathbf{X}}^H(n)\widetilde{\mathbf{X}}(n)\right]^{-1}\mathbf{D}_\alpha(n)\mathbf{e}^*(n), \tag{5.42}$$

where

$$\mathbf{D}_\alpha(n) = \mathrm{diag}\left[\alpha_0(n)\ \alpha_1(n)\ \cdots\ \alpha_{P-1}(n)\right] \tag{5.43}$$

is a $P \times P$ diagonal matrix. It is clear that (5.41) is obtained when $\alpha_0(n) = \alpha_1(n) = \cdots = \alpha_{P-1}(n) = \alpha$. Substituting (5.42) into (5.2), the posteriori error vector becomes

$$\boldsymbol{\epsilon}(n) = \left[\mathbf{I}_P - \mathbf{D}_\alpha(n)\right]\mathbf{e}(n). \tag{5.44}$$

In consistence with the basic idea of the APA, it can be imposed to cancel P a posteriori errors, i.e., $\boldsymbol{\epsilon}(n) = \mathbf{0}$. Assuming that $\mathbf{e}(n) \neq \mathbf{0}$, it results from (5.44) that $\mathbf{D}_\alpha(n) = \mathbf{I}_P$. This corresponds to the update (5.41), with the step size $\alpha = 1$. In the absence of the system noise, i.e., $v(n) = 0$, we deal with an ideal "system identification" configuration. In this case, the value of the step size $\alpha = 1$ makes sense, because it leads to the best performance [1]. However, in most of real-world system identification scenarios (e.g., like echo cancellation) the existence of the system noise cannot be omitted, so that a more reasonable condition to impose is $\boldsymbol{\epsilon}(n) = \mathbf{v}(n)$, i.e., to recover the system noise from the error of the adaptive filter. Thus, taking (5.44) into account, it results that

$$\begin{aligned}\epsilon_p(n) &= \left[1 - \alpha_p(n)\right]e_p(n)\\ &= v(n-p),\end{aligned} \tag{5.45}$$

where $\epsilon_p(n)$ and $e_p(n)$ denote the $(p+1)$th elements of the vectors $\boldsymbol{\epsilon}(n)$ and $\mathbf{e}(n)$, with $p = 0, 1, \ldots, P-1$. Our goal is to find an expression for the normalized step-size parameter $\alpha_p(n)$ such that

$$E\left[|\epsilon_p(n)|^2\right] = E\left[|v(n-p)|^2\right]. \tag{5.46}$$

From (5.45), we deduce that

$$\left[1 - \alpha_p(n)\right]^2 E\left[|e_p(n)|^2\right] = E\left[|v(n-p)|^2\right]. \tag{5.47}$$

By solving the quadratic equation (5.47), two solutions can be obtained. However, a value of the step size between 0 and 1 is preferable over the one between 1 and 2 (even if both solutions are stable but the former has less steady-state MSE with the same convergence speed [6]), so that it is reasonable to choose

$$\alpha_p(n) = 1 - \sqrt{\frac{E\left[|v(n-p)|^2\right]}{E\left[|e_p(n)|^2\right]}}. \tag{5.48}$$

From a practical point of view, (5.48) has to be evaluated in terms of power estimates as

$$\alpha_p(n) = 1 - \frac{\widehat{\sigma}_v(n-p)}{\widehat{\sigma}_{e_p}(n)}. \tag{5.49}$$

The variable in the denominator can be computed in a recursive manner (see Chapter 4), i.e.,

$$\widehat{\sigma}^2_{e_p}(n) = \lambda_{\mathrm{w}}\widehat{\sigma}^2_{e_p}(n-1) + (1-\lambda_{\mathrm{w}})\,|e_p(n)|^2. \tag{5.50}$$

However, the main problem remains the estimation of the system noise power from the numerator of (5.49). In order to analyze this aspect, we consider the echo cancellation scenario. First, it is known that in the single-talk case, the near-end signal consists only of the background noise. Its power could be estimated during silences (and it can be assumed constant), so that (5.49) becomes

$$\alpha_p(n) = 1 - \frac{\sigma_v}{\widehat{\sigma}_{e_p}(n)}. \tag{5.51}$$

For a value of the projection order $P = 1$, the VSS-NLMS algorithm from Section 4.6, Chapter 4 is obtained. For $P > 1$, a VSS-APA can be derived, by computing (5.51) for $p = 0, 1, \ldots, P-1$, then using a step-size matrix like in (5.43), and updating the filter coefficients according to (5.42). Nevertheless, the background noise can be time-variant, so that the power of the background noise should be periodically estimated. Moreover, when the background noise changes between two consecutive estimations or during the near-end speech, its new power estimate will not be available immediately; consequently, until the next estimation period of the background noise, the algorithm behavior will be disturbed. Second, in the double-talk case, the near-end signal consists of both the background noise and the near-end speech. It is very difficult to obtain an accurate estimate for the power of this combined signal, considering especially the nonstationary character of the speech signal.

In order to overcome these issues, let us consider the approach proposed in [7], which provides a simple but practical way to evaluate the numerator in (5.49). We recall that the complex observation (output) signal can be expressed as

$$d(n) = y(n) + v(n). \tag{5.52}$$

We deduce that

$$E\left[|v(n)|^2\right] = E\left[|d(n)|^2\right] - E\left[|y(n)|^2\right]. \tag{5.53}$$

Assuming that the adaptive filter has converged to a certain degree, it can be considered that

$$E\left[|y(n)|^2\right] \approx E\left[|\widehat{y}(n)|^2\right], \tag{5.54}$$

where $\widehat{y}(n) = \widetilde{\mathbf{h}}^H(n-1)\widetilde{\mathbf{x}}(n)$ is the output of the adaptive filter. Consequently,

$$E\left[|v(n)|^2\right] \approx E\left[|d(n)|^2\right] - E\left[|\widehat{y}(n)|^2\right] \tag{5.55}$$

or in terms of power estimates,

$$\widehat{\sigma}_v^2(n) \approx \widehat{\sigma}_d^2(n) - \widehat{\sigma}_{\widehat{y}}^2(n). \tag{5.56}$$

For the single-talk case, when only the background noise is present at the near-end, an estimate of its power is obtained using the right-hand term in (5.56). This expression holds even if the level of the background noise changes, so that there is no need for the estimation of this parameter during silences. For the double-talk case, when the near-end speech is present (assuming that it is uncorrelated with the background noise), the right-hand term in (5.56) also provides a power estimate of the near-end signal. More importantly, this term depends only on the signals that are available within the application, i.e., the system output (observation) signal, $d(n)$, and the adaptive filter output, $\widehat{y}(n)$. Based on these findings, (5.49) can be rewritten as

$$\alpha_p(n) = 1 - \frac{\sqrt{\widehat{\sigma}_d^2(n-p) - \widehat{\sigma}_{\widehat{y}}^2(n-p)}}{\widehat{\sigma}_{e_p}(n)}, \quad p = 0, 1, \ldots, P-1. \tag{5.57}$$

As compared to (5.51), the previous relation is more suitable in practice. It should be noted that both terms from the numerator on the right-hand side of (5.57) can be evaluated using a recursive procedure similar to (5.50).

Under our assumptions, we have $E\left[|d(n-p)|^2\right] \geq E\left[|\widehat{y}(n-p)|^2\right]$ and $E\left[|d(n-p)|^2\right] - E\left[|\widehat{y}(n-p)|^2\right] \approx E\left[|e_p(n)|^2\right]$. Nevertheless, the power estimates of these parameters could lead to some deviations from the previous theoretical conditions, so that we will take the absolute values in (5.57). Hence, the final step-size formula is rewritten as

$$\alpha_p(n) = \left| 1 - \frac{\sqrt{\left| \widehat{\sigma}_d^2(n-p) - \widehat{\sigma}_{\widehat{y}}^2(n-p) \right|}}{\widehat{\sigma}_{e_p}(n)} \right|, \ p = 0, 1, \ldots, P-1. \quad (5.58)$$

Finally, the update equation of the VSS-APA is

$$\widetilde{\mathbf{h}}(n) = \widetilde{\mathbf{h}}(n-1) + \widetilde{\mathbf{X}}(n) \left[\widetilde{\mathbf{X}}^H(n)\widetilde{\mathbf{X}}(n) + \delta \mathbf{I}_P \right]^{-1} \mathbf{D}_\alpha(n)\mathbf{e}^*(n), \quad (5.59)$$

where the regularization parameter, δ, should be the same as for the APA and the elements of the diagonal matrix $\mathbf{D}_\alpha(n)$ are defined in (5.58).

5.5 Improved Proportionate APA (IPAPA)

As explained in Chapter 4, it makes more sense to use the minimum ℓ_1-norm solution than the minimum ℓ_2-norm solution in any type of adaptive filters when the impulse response is sparse.

We now derive a proportionate-type APA following the steps of our interpretation of the APA [3].

We start by solving the optimization problem

$$\min_{\overleftarrow{\mathbf{h}}(n)} \left\| \overleftarrow{\mathbf{h}}(n) \right\|_1 \quad \text{subject to} \quad \mathbf{d}(n) = \widetilde{\mathbf{X}}^T(n)\overleftarrow{\mathbf{h}}^*(n). \quad (5.60)$$

Using Lagrange multipliers, we find that

$$\overleftarrow{\mathbf{h}}(n) = \overleftarrow{\mathbf{G}}(n)\widetilde{\mathbf{X}}(n) \left[\widetilde{\mathbf{X}}^H(n)\overleftarrow{\mathbf{G}}(n)\widetilde{\mathbf{X}}(n) \right]^{-1} \mathbf{d}^*(n), \quad (5.61)$$

where $\overleftarrow{\mathbf{G}}(n)$ is defined in (4.87). Since (5.61) is hard to solve, it can be well approximated by

$$\overleftarrow{\mathbf{h}}(n) = \mathbf{G}(n-1)\widetilde{\mathbf{X}}(n) \left[\widetilde{\mathbf{X}}^H(n)\mathbf{G}(n-1)\widetilde{\mathbf{X}}(n) \right]^{-1} \mathbf{d}^*(n), \quad (5.62)$$

where $\mathbf{G}(n-1)$ is defined in (4.88).

We then deduce a proportionate-type APA:

$$\widetilde{\mathbf{h}}(n) = \mathbf{P}(n)\widetilde{\mathbf{h}}(n-1) + \overleftarrow{\mathbf{h}}(n), \quad (5.63)$$

where

$$\mathbf{P}(n) = \mathbf{I}_{2L} - \mathbf{G}(n-1)\widetilde{\mathbf{X}}(n) \left[\widetilde{\mathbf{X}}^H(n)\mathbf{G}(n-1)\widetilde{\mathbf{X}}(n) \right]^{-1} \widetilde{\mathbf{X}}^H(n) \quad (5.64)$$

is a (non-orthogonal) projection matrix. Expression (5.63) can be rewritten as

$$\widetilde{\mathbf{h}}(n) = \widetilde{\mathbf{h}}(n-1) + \mathbf{G}(n-1)\widetilde{\mathbf{X}}(n)\left[\widetilde{\mathbf{X}}^H(n)\mathbf{G}(n-1)\widetilde{\mathbf{X}}(n)\right]^{-1}\mathbf{e}^*(n). \quad (5.65)$$

Making this algorithm more practical to avoid problems such as stalling of the coefficients, will lead to the well-known proportionate APA (PAPA) and improved APA (IPAPA) [8], [9].

To make (5.65) more practical, we propose to write it as an IPAPA form [10], i.e.,

$$\widetilde{\mathbf{h}}(n) = \widetilde{\mathbf{h}}(n-1) + \alpha\mathbf{G}(n-1)\widetilde{\mathbf{X}}(n)\left[\widetilde{\mathbf{X}}^H(n)\mathbf{G}(n-1)\widetilde{\mathbf{X}}(n) + \delta\mathbf{I}_P\right]^{-1}\mathbf{e}^*(n), \quad (5.66)$$

where α $(0 < \alpha < 2)$ is the normalized step-size parameter, the elements of the diagonal matrix $\mathbf{G}(n-1)$ are defined in (4.95), and the regularization parameter is found to be

$$\delta = \beta_{\text{IPAPA}}\sigma_x^2, \quad (5.67)$$

where

$$\beta_{\text{IPAPA}} = \frac{1 + \sqrt{1 + \text{SENR}}}{\text{SENR}} \quad (5.68)$$

is the normalized regularization parameter of the IPAPA.

We can also easily deduce the update equation of the VSS-IPAPA:

$$\begin{aligned}\widetilde{\mathbf{h}}(n) &= \widetilde{\mathbf{h}}(n-1) \quad (5.69)\\ &+ \mathbf{G}(n-1)\widetilde{\mathbf{X}}(n)\left[\widetilde{\mathbf{X}}^H(n)\mathbf{G}(n-1)\widetilde{\mathbf{X}}(n) + \delta\mathbf{I}_P\right]^{-1}\mathbf{D}_\alpha(n)\mathbf{e}^*(n),\end{aligned}$$

where the elements of the diagonal matrix $\mathbf{D}_\alpha(n)$ are defined in (5.58).

5.6 Memory PAPA

The update equation of the PAPA can be expressed as

$$\widetilde{\mathbf{h}}(n) = \widetilde{\mathbf{h}}(n-1) + \alpha\widetilde{\mathbf{X}}_{\mathbf{g}}(n)\left[\widetilde{\mathbf{X}}^H(n)\widetilde{\mathbf{X}}_{\mathbf{g}}(n) + \delta\mathbf{I}_P\right]^{-1}\mathbf{e}^*(n), \quad (5.70)$$

where

$$\widetilde{\mathbf{X}}_{\mathbf{g}}(n) = \mathbf{G}(n-1)\widetilde{\mathbf{X}}(n) \quad (5.71)$$

and

$$\mathbf{G}(n-1) = \mathrm{diag}\left[g_0(n-1)\; g_1(n-1)\; \cdots\; g_{2L-1}(n-1)\right] \tag{5.72}$$

is a $2L \times 2L$ diagonal matrix that assigns an individual step size to each filter coefficient (in order to "proportionate" the algorithm behavior). We can rewrite (5.71) as

$$\widetilde{\mathbf{X}}_{\mathbf{g}}(n) = \left[\mathbf{g}(n-1) \odot \widetilde{\mathbf{x}}(n)\; \cdots\; \mathbf{g}(n-1) \odot \widetilde{\mathbf{x}}(n-P+1)\right], \tag{5.73}$$

where

$$\mathbf{g}(n-1) = \left[g_0(n-1)\; g_1(n-1)\; \cdots\; g_{2L-1}(n-1)\right]^T$$

is a vector containing the diagonal elements of $\mathbf{G}(n-1)$ and the operator $\odot$ denotes the Hadamard product. Clearly, when implementing the PAPA in practice, (5.73) is used, requiring $2PL$ complex multiplications for evaluating the matrix $\widetilde{\mathbf{X}}_{\mathbf{g}}(n)$.

Nevertheless, the APA can be viewed as an algorithm with "memory," i.e., it takes into account the "history" of the last P time samples. The classical PAPA does not take into account the "proportionate history" of each coefficient $\widetilde{h}_l(n-1)$, $l = 0, 1, \ldots, 2L-1$, but only its proportionate factor from the current time sample, i.e., $g_l\,(n-1)$. Therefore, the recently proposed memory PAPA (MPAPA) [11] takes advantage of the "proportionate memory" of the algorithm, by choosing the matrix

$$\widetilde{\mathbf{X}}'_{\mathbf{g}}(n) = \left[\mathbf{g}(n-1) \odot \widetilde{\mathbf{x}}(n)\; \cdots\; \mathbf{g}(n-P) \odot \widetilde{\mathbf{x}}(n-P+1)\right], \tag{5.74}$$

instead of the matrix $\widetilde{\mathbf{X}}_{\mathbf{g}}(n)$ from (5.73). In this manner, we take into account the "proportionate history" of the coefficient $\widetilde{h}_l(n-1)$, in terms of its proportionate factors from the last P time samples.

The advantage of this modification is twofold. First, the MPAPA takes into account the "history" of the proportionate factors from the last P steps. Of course the gain in terms of fast convergence and tracking becomes more apparent when the projection order P increases. Second, the computational complexity is lower as compared to the classical PAPA. This is because (5.74) can be implemented recursively as

$$\widetilde{\mathbf{X}}'_{\mathbf{g}}(n) = \left[\mathbf{g}(n-1) \odot \widetilde{\mathbf{x}}(n)\; \widetilde{\mathbf{X}}'_{\mathbf{g},-1}(n-1)\right], \tag{5.75}$$

where the matrix

$$\widetilde{\mathbf{X}}'_{\mathbf{g},-1}(n-1) = \left[\mathbf{g}(n-2) \odot \widetilde{\mathbf{x}}(n-1)\; \cdots\; \mathbf{g}(n-P) \odot \widetilde{\mathbf{x}}(n-P+1)\right]$$

contains the first $P-1$ columns of $\widetilde{\mathbf{X}}'_{\mathbf{g}}(n-1)$. Thus, the columns from 1 to $P-1$ of the matrix $\widetilde{\mathbf{X}}'_{\mathbf{g}}(n-1)$ can be used directly for computing the matrix

$\widetilde{\mathbf{X}}'_{\mathbf{g}}(n)$. This is not the case in the classical PAPAs, where all the columns of $\widetilde{\mathbf{X}}_{\mathbf{g}}(n)$ [see (5.73)] have to be evaluated at each iteration, because all of them are multiplied with the same vector $\mathbf{g}(n-1)$. Consequently, the evaluation of $\widetilde{\mathbf{X}}_{\mathbf{g}}(n)$ from (5.73) needs $2PL$ complex multiplications, while the evaluation of $\widetilde{\mathbf{X}}'_{\mathbf{g}}(n)$ [see (5.75)] requires only $2L$ complex multiplications. Clearly, this advantage becomes more apparent when the projection order P increases. Concluding, the MPAPA is more computationally efficient as compared to the classical PAPAs. Clearly, we can also derive a VSS-MPAPA as in [12].

References

1. K. Ozeki and T. Umeda, "An adaptive filtering algorithm using an orthogonal projection to an affine subspace and its properties," *Electron. Commun. Jpn.*, vol. 67-A, pp. 19–27, May 1984.
2. Y. Xia, C. Cheong Took, and D. P. Mandic, "An augmented affine projection algorithm for the filtering of noncircular complex signals," *Signal Process.*, vol. 90, pp. 1788–1799, June 2010.
3. J. Benesty, C. Paleologu, and S. Ciochină, "Proportionate adaptive filters from a basis pursuit perspective," *IEEE Signal Process. Lett.*, vol. 17, pp. 985–988, Dec. 2010.
4. C. Paleologu, J. Benesty, and S. Ciochină, "Regularization of the affine projection algorithm," *IEEE Trans. Circuits and Systems II: Express Briefs*, to appear, 2011.
5. W. Yin and A. S. Mehr, "A variable regularization method for affine projection algorithm," *IEEE Trans. Circuits and Systems II: Express Briefs*, vol. 57, pp. 476–480, June 2010.
6. S. G. Sankaran and A. A. L. Beex, "Convergence behavior of affine projection algorithms," *IEEE Trans. Signal Process.*, vol. 48, pp. 1086–1096, Apr. 2000.
7. C. Paleologu, J. Benesty, and S. Ciochină, "A variable step-size affine projection algorithm designed for acoustic echo cancellation," *IEEE Trans. Audio, Speech, Language Process.*, vol. 16, pp. 1466–1478, Nov. 2008.
8. T. Gaensler, S. L. Gay, M. M. Sondhi, and J. Benesty, "Double-talk robust fast converging algorithms for network echo cancellation," *IEEE Trans. Speech, Audio Process.*, vol. 8, pp. 656–663, Nov. 2000.
9. O. Hoshuyama, R. A. Goubran, and A. Sugiyama, "A generalized proportionate variable step-size algorithm for fast changing acoustic environments," in *Proc. IEEE ICASSP*, 2004, pp. IV-161–IV-164.
10. J. Benesty and S. L. Gay, "An improved PNLMS algorithm," in *Proc. IEEE ICASSP*, 2002, pp. 1881–1884.
11. C. Paleologu, S. Ciochină, and J. Benesty, "An efficient proportionate affine projection algorithm for echo cancellation," *IEEE Signal Process. Lett.*, vol. 17, pp. 165–168, Feb. 2010.
12. C. Paleologu, J. Benesty, F. Albu, and S. Ciochină, "An efficient variable step-size proportionate affine projection algorithm," in *Proc. IEEE ICASSP*, 2011, pp. 77–80.

Chapter 6
Recursive Least-Squares Algorithms

Thanks to their fast convergence rate, recursive least-squares (RLS) algorithms are very popular in SAEC [1]. Indeed, it is well known that the convergence rate of RLS-type algorithms are not much affected by the nature of the input signal, even when this one is ill-conditioned. Actually, the very first SAEC prototype was based on the fast RLS (FRLS) algorithm, which was implemented in subbands [2]. In this chapter, we derive the RLS and FRLS algorithms in the WL context.

6.1 Least-Squares Error Criterion and Normal Equations

In this chapter only, we slightly change the notation for convenience. We redefine the input signal vector (of length $2L$) as

$$\widetilde{\mathbf{x}}(n) = \left[\boldsymbol{\chi}^T(n) \ \boldsymbol{\chi}^T(n-1) \cdots \boldsymbol{\chi}^T(n-L+1) \right]^T, \qquad (6.1)$$

where

$$\boldsymbol{\chi}(n) = \left[x(n) \ x^*(n) \right]^T. \qquad (6.2)$$

As a result, the new definitions of the true impulse response and the adaptive filter are

$$\widetilde{\mathbf{h}}_{\mathrm{t}} = \left[h_{\mathrm{t},0} \ h'_{\mathrm{t},0} \cdots h_{\mathrm{t},L-1} \ h'_{\mathrm{t},L-1} \right]^T, \qquad (6.3)$$

$$\widetilde{\mathbf{h}}(n) = \left[h_0(n) \ h'_0(n) \cdots h_{L-1}(n) \ h'_{L-1}(n) \right]^T, \qquad (6.4)$$

where $h_{\mathrm{t},l}$, $h'_{\mathrm{t},l}$, $h_l(n)$, and $h'_l(n)$, with $l = 0, 1, \ldots, L-1$, are the elements of the vectors $\mathbf{h}_{\mathrm{t}}$, $\mathbf{h}'_{\mathrm{t}}$, $\mathbf{h}(n)$, and $\mathbf{h}'(n)$, respectively. We recommend the reader to go back to Chapter 2 to see the differences (i.e., the vectors are interleaved

J. Benesty et al.: A Perspective on Stereophonic Acoustic Echo Cancellation, STSP 4, pp. 63–69.
springerlink.com

now instead of being concatenated). Obviously, these new definitions do not change the definition of the complex observation, which is

$$d(n) = \widetilde{\mathbf{h}}_{\mathrm{t}}^{H}\widetilde{\mathbf{x}}(n) + v(n). \tag{6.5}$$

We define the least-squares (LS) error criterion as [3]

$$J\left[\widetilde{\mathbf{h}}(n)\right] = \sum_{i=1}^{n} \lambda_{\mathrm{L}}^{n-i} \left| d(i) - \widetilde{\mathbf{h}}^{H}(n)\widetilde{\mathbf{x}}(i)\right|^{2}, \tag{6.6}$$

where λ_{L} $(0 \ll \lambda_{\mathrm{L}} < 1)$ is the forgetting factor, which influences the memory of the data in the different statistics estimates. The special case of $\lambda_{\mathrm{L}} = 1$ corresponds to infinite memory. We can express (6.6) as

$$J\left[\widetilde{\mathbf{h}}(n)\right] = \sigma_d^2(n) - \widetilde{\mathbf{h}}^{H}(n)\mathbf{p}_{\widetilde{\mathbf{x}}d}(n) - \mathbf{p}_{\widetilde{\mathbf{x}}d}^{H}(n)\widetilde{\mathbf{h}}(n) + \widetilde{\mathbf{h}}^{H}(n)\mathbf{R}_{\widetilde{\mathbf{x}}}(n)\widetilde{\mathbf{h}}(n), \tag{6.7}$$

where

$$\sigma_d^2(n) = \sum_{i=1}^{n} \lambda_{\mathrm{L}}^{n-i} \left| d(i)\right|^2, \tag{6.8}$$

$$\mathbf{p}_{\widetilde{\mathbf{x}}d}(n) = \sum_{i=1}^{n} \lambda_{\mathrm{L}}^{n-i}\widetilde{\mathbf{x}}(i)d^{*}(i), \tag{6.9}$$

$$\mathbf{R}_{\widetilde{\mathbf{x}}}(n) = \sum_{i=1}^{n} \lambda_{\mathrm{L}}^{n-i}\widetilde{\mathbf{x}}(i)\widetilde{\mathbf{x}}^{H}(i). \tag{6.10}$$

The minimization of $J\left[\widetilde{\mathbf{h}}(n)\right]$ with respect to $\widetilde{\mathbf{h}}(n)$ leads to the well-known normal equations [3], [4]:

$$\mathbf{R}_{\widetilde{\mathbf{x}}}(n)\widetilde{\mathbf{h}}(n) = \mathbf{p}_{\widetilde{\mathbf{x}}d}(n). \tag{6.11}$$

Assuming that $\mathbf{R}_{\widetilde{\mathbf{x}}}(n) > 0$, we deduce that the optimal filter in the LS sense is

$$\widetilde{\mathbf{h}}(n) = \mathbf{R}_{\widetilde{\mathbf{x}}}^{-1}(n)\mathbf{p}_{\widetilde{\mathbf{x}}d}(n). \tag{6.12}$$

Solving the previous equation with classical approaches such as the Gaussian elimination will require an arithmetic complexity proportional to $(2L)^3$. However, by just taking into account the recursions of the different variables, we can reduce this complexity by a factor of $2L$ as explained in the next section.

6.2 Recursive Least-Squares (RLS) Algorithm

It is easy to see from (6.9) and (6.10) that $\mathbf{p}_{\widetilde{\mathbf{x}}d}(n)$ and $\mathbf{R}_{\widetilde{\mathbf{x}}}(n)$ can be computed recursively as follows:

$$\mathbf{p}_{\widetilde{\mathbf{x}}d}(n) = \lambda_{\mathrm{L}}\mathbf{p}_{\widetilde{\mathbf{x}}d}(n-1) + \widetilde{\mathbf{x}}(n)d^*(n), \tag{6.13}$$

$$\mathbf{R}_{\widetilde{\mathbf{x}}}(n) = \lambda_{\mathrm{L}}\mathbf{R}_{\widetilde{\mathbf{x}}}(n-1) + \widetilde{\mathbf{x}}(n)\widetilde{\mathbf{x}}^H(n). \tag{6.14}$$

Applying the Woodbury's identity in (6.14), we find that the inverse of $\mathbf{R}_{\widetilde{\mathbf{x}}}(n)$ can be expressed as

$$\begin{aligned}\mathbf{R}_{\widetilde{\mathbf{x}}}^{-1}(n) &= \lambda_{\mathrm{L}}^{-1}\mathbf{R}_{\widetilde{\mathbf{x}}}^{-1}(n-1) - \frac{\lambda_{\mathrm{L}}^{-2}\mathbf{R}_{\widetilde{\mathbf{x}}}^{-1}(n-1)\widetilde{\mathbf{x}}(n)\widetilde{\mathbf{x}}^H(n)\mathbf{R}_{\widetilde{\mathbf{x}}}^{-1}(n-1)}{1+\lambda_{\mathrm{L}}^{-1}\widetilde{\mathbf{x}}^H(n)\mathbf{R}_{\widetilde{\mathbf{x}}}^{-1}(n-1)\widetilde{\mathbf{x}}(n)} \\ &= \lambda_{\mathrm{L}}^{-1}\mathbf{R}_{\widetilde{\mathbf{x}}}^{-1}(n-1) - \lambda_{\mathrm{L}}^{-1}\widetilde{\mathbf{k}}(n)\widetilde{\mathbf{x}}^H(n)\mathbf{R}_{\widetilde{\mathbf{x}}}^{-1}(n-1),\end{aligned} \tag{6.15}$$

where

$$\widetilde{\mathbf{k}}(n) = \frac{\lambda_{\mathrm{L}}^{-1}\mathbf{R}_{\widetilde{\mathbf{x}}}^{-1}(n-1)\widetilde{\mathbf{x}}(n)}{1+\lambda_{\mathrm{L}}^{-1}\widetilde{\mathbf{x}}^H(n)\mathbf{R}_{\widetilde{\mathbf{x}}}^{-1}(n-1)\widetilde{\mathbf{x}}(n)} \tag{6.16}$$

is the Kalman gain vector. Rearranging the previous equation, we obtain

$$\begin{aligned}\widetilde{\mathbf{k}}(n) &= \lambda_{\mathrm{L}}^{-1}\mathbf{R}_{\widetilde{\mathbf{x}}}^{-1}(n-1)\widetilde{\mathbf{x}}(n) - \lambda_{\mathrm{L}}^{-1}\widetilde{\mathbf{k}}(n)\widetilde{\mathbf{x}}^H(n)\mathbf{R}_{\widetilde{\mathbf{x}}}^{-1}(n-1)\widetilde{\mathbf{x}}(n) \\ &= \left[\lambda_{\mathrm{L}}^{-1}\mathbf{R}_{\widetilde{\mathbf{x}}}^{-1}(n-1) - \lambda_{\mathrm{L}}^{-1}\widetilde{\mathbf{k}}(n)\widetilde{\mathbf{x}}^H(n)\mathbf{R}_{\widetilde{\mathbf{x}}}^{-1}(n-1)\right]\widetilde{\mathbf{x}}(n) \\ &= \mathbf{R}_{\widetilde{\mathbf{x}}}^{-1}(n)\widetilde{\mathbf{x}}(n).\end{aligned} \tag{6.17}$$

Now, we can rewrite (6.12) as

$$\widetilde{\mathbf{h}}(n) = \lambda_{\mathrm{L}}\mathbf{R}_{\widetilde{\mathbf{x}}}^{-1}(n)\mathbf{p}_{\widetilde{\mathbf{x}}d}(n-1) + \mathbf{R}_{\widetilde{\mathbf{x}}}^{-1}(n)\widetilde{\mathbf{x}}(n)d^*(n). \tag{6.18}$$

Substituting (6.15) into the first term only on the right-hand side of (6.18), we obtain

$$\begin{aligned}\widetilde{\mathbf{h}}(n) &= \mathbf{R}_{\widetilde{\mathbf{x}}}^{-1}(n-1)\mathbf{p}_{\widetilde{\mathbf{x}}d}(n-1) - \widetilde{\mathbf{k}}(n)\widetilde{\mathbf{x}}^H(n)\mathbf{R}_{\widetilde{\mathbf{x}}}^{-1}(n-1)\mathbf{p}_{\widetilde{\mathbf{x}}d}(n-1) \\ &\quad + \mathbf{R}_{\widetilde{\mathbf{x}}}^{-1}(n)\widetilde{\mathbf{x}}(n)d^*(n) \\ &= \widetilde{\mathbf{h}}(n-1) - \widetilde{\mathbf{k}}(n)\widetilde{\mathbf{x}}^H(n)\widetilde{\mathbf{h}}(n-1) + \widetilde{\mathbf{k}}(n)d^*(n) \\ &= \widetilde{\mathbf{h}}(n-1) + \widetilde{\mathbf{k}}(n)\left[d^*(n) - \widetilde{\mathbf{x}}^H(n)\widetilde{\mathbf{h}}(n-1)\right] \\ &= \widetilde{\mathbf{h}}(n-1) + \widetilde{\mathbf{k}}(n)e^*(n),\end{aligned} \tag{6.19}$$

where

Table 6.1 RLS algorithm.

Initialization:
$\widetilde{\mathbf{h}}(0) = \mathbf{0}$
$\mathbf{R}_{\widetilde{\mathbf{x}}}^{-1}(0) = \delta^{-1}\mathbf{I}_{2L}$

Parameters:
$\lambda_{\mathrm{L}} = 1 - \frac{1}{2C_{\mathrm{L}}L}$, forgetting factor with $C_{\mathrm{L}} \geq 3$
$\delta > 0$, regularization

For time index $n = 1, 2, \ldots$:
$\widetilde{\mathbf{k}}(n) = \frac{\mathbf{R}_{\widetilde{\mathbf{x}}}^{-1}(n-1)\widetilde{\mathbf{x}}(n)}{\lambda_{\mathrm{L}} + \widetilde{\mathbf{x}}^H(n)\mathbf{R}_{\widetilde{\mathbf{x}}}^{-1}(n-1)\widetilde{\mathbf{x}}(n)}$
$e(n) = d(n) - \widetilde{\mathbf{h}}^H(n-1)\widetilde{\mathbf{x}}(n)$
$\widetilde{\mathbf{h}}(n) = \widetilde{\mathbf{h}}(n-1) + \widetilde{\mathbf{k}}(n)e^*(n)$
$\mathbf{R}_{\widetilde{\mathbf{x}}}^{-1}(n) = \lambda_{\mathrm{L}}^{-1}\mathbf{R}_{\widetilde{\mathbf{x}}}^{-1}(n-1) - \lambda_{\mathrm{L}}^{-1}\widetilde{\mathbf{k}}(n)\widetilde{\mathbf{x}}^H(n)\mathbf{R}_{\widetilde{\mathbf{x}}}^{-1}(n-1)$

$$e(n) = d(n) - \widetilde{\mathbf{h}}^H(n-1)\widetilde{\mathbf{x}}(n) \tag{6.20}$$

is the a priori error signal.

The a posteriori error signal is

$$\begin{aligned}\epsilon(n) &= d(n) - \widetilde{\mathbf{h}}^H(n)\widetilde{\mathbf{x}}(n) \\ &= \varphi(n)e(n),\end{aligned} \tag{6.21}$$

where

$$\varphi(n) = 1 - \widetilde{\mathbf{x}}^H(n)\mathbf{R}_{\widetilde{\mathbf{x}}}^{-1}(n)\widetilde{\mathbf{x}}(n). \tag{6.22}$$

It is not hard to show that

$$0 < \varphi(n) < 1. \tag{6.23}$$

As a consequence,

$$|\epsilon(n)| \leq |e(n)|. \tag{6.24}$$

Equations (6.15), (6.16), (6.19), and (6.20) constitute the RLS algorithm, which is summarized in Table 6.1 [3], [4]. It can be checked that now the arithmetic complexity is proportional to $(2L)^2$.

6.3 Fast RLS (FRLS) Algorithm

The WL RLS developed in the preceding section can be seen as a two-channel RLS algorithm, since $x(n)$ and $x^*(n)$ correspond to the inputs of these two channels. Therefore, we can follow the steps shown in [3] to derive a fast RLS (FRLS) algorithm, whose arithmetic complexity is proportional to $2L$, thanks to the different relations between the forward and backward predictors.

We derive a fast version based on the a priori Kalman gain vector defined as

$$\widetilde{\mathbf{k}}'(n) = \mathbf{R}_{\widetilde{\mathbf{x}}}^{-1}(n-1)\widetilde{\mathbf{x}}(n). \tag{6.25}$$

With this definition, the update equation of the RLS algorithm becomes

$$\widetilde{\mathbf{h}}(n) = \widetilde{\mathbf{h}}(n-1) + \widetilde{\mathbf{k}}'(n)\frac{e^*(n)}{\theta(n)}, \tag{6.26}$$

where

$$\begin{aligned} \theta(n) &= \widetilde{\mathbf{x}}^H(n)\widetilde{\mathbf{k}}'(n) \\ &= \frac{\lambda_{\mathrm{L}}}{\varphi(n)}. \end{aligned} \tag{6.27}$$

We define the forward and backward prediction error energy matrices as

$$\begin{aligned} \mathbf{E}_{\mathrm{f}}(n) = \sum_{i=1}^{n} \lambda_{\mathrm{L}}^{n-i} \left[\boldsymbol{\chi}(i) - \mathbf{A}^H(n-1)\widetilde{\mathbf{x}}(i-1)\right] \times \\ \left[\boldsymbol{\chi}(i) - \mathbf{A}^H(n-1)\widetilde{\mathbf{x}}(i-1)\right]^H, \end{aligned} \tag{6.28}$$

$$\begin{aligned} \mathbf{E}_{\mathrm{b}}(n) = \sum_{i=1}^{n} \lambda_{\mathrm{L}}^{n-i} \left[\boldsymbol{\chi}(i-L) - \mathbf{B}^H(n-1)\widetilde{\mathbf{x}}(i)\right] \times \\ \left[\boldsymbol{\chi}(i-L) - \mathbf{B}^H(n-1)\widetilde{\mathbf{x}}(i)\right]^H, \end{aligned} \tag{6.29}$$

where $\mathbf{A}(n-1)$ and $\mathbf{B}(n-1)$ are the forward and backward coefficient matrices of size $2L \times 2$. The minimization of $\mathrm{tr}\left[\mathbf{E}_{\mathrm{f}}(n)\right]$ and $\mathrm{tr}\left[\mathbf{E}_{\mathrm{b}}(n)\right]$ leads to

$$\mathbf{A}(n) = \mathbf{A}(n-1) + \widetilde{\mathbf{k}}'(n-1)\frac{\mathbf{e}_{\mathrm{f}}^H(n)}{\theta(n-1)}, \tag{6.30}$$

$$\mathbf{B}(n) = \mathbf{B}(n-1) + \widetilde{\mathbf{k}}'(n)\frac{\mathbf{e}_{\mathrm{b}}^H(n)}{\theta(n)}, \tag{6.31}$$

where

$$\mathbf{e}_{\mathrm{f}}(n) = \boldsymbol{\chi}(n) - \mathbf{A}^H(n-1)\widetilde{\mathbf{x}}(n-1) \tag{6.32}$$

Table 6.2 FRLS algorithm.

Prediction:

$$\mathbf{e}_\mathrm{f}(n) = \boldsymbol{\chi}(n) - \mathbf{A}^H(n-1)\widetilde{\mathbf{x}}(n-1)$$

$$\theta_1(n) = \theta(n-1) + \mathbf{e}_\mathrm{f}^H(n)\mathbf{E}_\mathrm{f}^{-1}(n-1)\mathbf{e}_\mathrm{f}(n)$$

$$\begin{bmatrix} \mathbf{k}(n) \\ \mathbf{j}(n) \end{bmatrix} = \begin{bmatrix} \mathbf{0} \\ \widetilde{\mathbf{k}}'(n-1) \end{bmatrix} + \begin{bmatrix} \mathbf{I}_2 \\ -\mathbf{A}(n-1) \end{bmatrix} \mathbf{E}_\mathrm{f}^{-1}(n-1)\mathbf{e}_\mathrm{f}(n)$$

$$\mathbf{E}_\mathrm{f}(n) = \lambda_\mathrm{L}\left[\mathbf{E}_\mathrm{f}(n-1) + \frac{\mathbf{e}_\mathrm{f}(n)\mathbf{e}_\mathrm{f}^H(n)}{\theta(n-1)}\right]$$

$$\mathbf{A}(n) = \mathbf{A}(n-1) + \widetilde{\mathbf{k}}'(n-1)\frac{\mathbf{e}_\mathrm{f}^H(n)}{\theta(n-1)}$$

$$\mathbf{e}_{\mathrm{b},1}(n) = \mathbf{E}_\mathrm{b}(n-1)\mathbf{j}(n)$$

$$\mathbf{e}_{\mathrm{b},2}(n) = \boldsymbol{\chi}(n-L) - \mathbf{B}^H(n-1)\widetilde{\mathbf{x}}(n)$$

$$\mathbf{e}_\mathrm{b}(n) = \kappa_\mathrm{s}\mathbf{e}_{\mathrm{b},2}(n) + (1-\kappa_\mathrm{s})\mathbf{e}_{\mathrm{b},1}(n)$$

$$\widetilde{\mathbf{k}}'(n) = \mathbf{k}(n) + \mathbf{B}(n-1)\mathbf{j}(n)$$

$$\theta(n) = \theta_1(n) - \mathbf{e}_{\mathrm{b},2}^H(n)\mathbf{j}(n)$$

$$\mathbf{E}_\mathrm{b}(n) = \lambda_\mathrm{L}\left[\mathbf{E}_\mathrm{b}(n-1) + \frac{\mathbf{e}_{\mathrm{b},2}(n)\mathbf{e}_{\mathrm{b},2}^H(n)}{\theta(n)}\right]$$

$$\mathbf{B}(n) = \mathbf{B}(n-1) + \widetilde{\mathbf{k}}'(n)\frac{\mathbf{e}_\mathrm{b}^H(n)}{\theta(n)}$$

Filtering:

$$e(n) = d(n) - \widetilde{\mathbf{h}}^H(n-1)\widetilde{\mathbf{x}}(n)$$

$$\widetilde{\mathbf{h}}(n) = \widetilde{\mathbf{h}}(n-1) + \widetilde{\mathbf{k}}'(n)\frac{e^*(n)}{\theta(n)}$$

$$\mathbf{e}_\mathrm{b}(n) = \boldsymbol{\chi}(n-L) - \mathbf{B}^H(n-1)\widetilde{\mathbf{x}}(n) \tag{6.33}$$

are the forward and backward prediction error vectors of length 2.

Exploiting all different relations, we derive the WL FRLS algorithm, which is summarized in Table 6.2 [3]. Note that there is another way to compute the backward prediction error vector. This form is taken into account in the table to "stabilize" the algorithm, where the stability parameter is denoted by κ_s ($1.5 \leq \kappa_\mathrm{s} \leq 2.5$). However, with nonstationary signals like speech, this version is not significantly more stable than its non-stabilized counterpart. Our approach to handle this problem, is simply to re-initialize the predictor-based variables when instability is detected with the use of the variable $\varphi(n)$, which is inherent in the FRLS computations. By monitoring $\varphi(n)$, whose values should always be between 0 and 1, it is possible to detect if the algorithm is about to become unstable. If this is the case, then the parameters in the prediction part are reset to their start values, while the adaptive filter estimate, $\widetilde{\mathbf{h}}(n)$, can be left unchanged. A suitable initial value for $\mathbf{A}(n)$, $\mathbf{B}(n)$, and $\widetilde{\mathbf{k}}'(n)$ are $\mathbf{0}$, whereas the energy estimates, $\mathbf{E}_{\mathrm{f}}(n)$ and $\mathbf{E}_{\mathrm{b}}(n)$ could be initialized with a recursive estimate of the speech energy.

Finally, the initialization of the FRLS algorithm should be as follows:

$$\begin{aligned}
\theta(0) &= \lambda_{\mathrm{L}}, \\
\mathbf{A}(0) &= \mathbf{B}(0) = \mathbf{0}, \\
\widetilde{\mathbf{k}}'(0) &= \mathbf{0}, \\
\mathbf{E}_{\mathrm{f}}(0) &\approx \frac{L\sigma_x^2}{100}\mathbf{I}_2, \\
\mathbf{E}_{\mathrm{b}}(0) &\approx \sigma_x^2 \lambda_{\mathrm{L}}^{-2L}\mathbf{I}_2.
\end{aligned}$$

References

1. J. Benesty, T. Gänsler, D. R. Morgan, M. M. Sondhi, and S. L. Gay, *Advances in Network and Acoustic Echo Cancellation*. Berlin, Germany: Springer-Verlag, 2001.
2. P. Eneroth, S. L. Gay, T. Gänsler, and J. Benesty, "A real-time implementation of a stereophonic acoustic echo canceler," *IEEE Trans. Audio, Speech Process.*, vol. 9, pp. 513–523, July 2001.
3. M. G. Bellanger, *Adaptive Filters and Signal Analysis*. NY: Dekker, 1988.
4. S. Haykin, *Adaptive Filter Theory*. Fourth Edition, Upper Saddle River, NJ: Prentice-Hall, 2002.

Chapter 7
Double-Talk Detection

Double-talk detectors (DTDs) are vital to the operation and performance of SAECs. In this chapter, we discuss the most well-known double-talk detection algorithms. It will be shown that, thanks to the WL formulation, DTDs for single-channel acoustic echo cancellation are easily generalized to the stereo case.

7.1 Principles of a Double-Talk Detector (DTD)

Ideally, SAECs remove the undesired echoes that result from the coupling between the two loudspeakers and the two microphones used in full-duplex hands-free stereo telecommunication systems. The (complex) far-end speech signal, $x(n)$, goes through the echo path represented by the complex filter $\widetilde{\mathbf{h}}_{\mathrm{t}}$, then it is picked up by the (complex) microphone together with the near-end talker signal $u(n)$ and ambient noise $v(n)$. The (complex) microphone signal is denoted as

$$d(n) = \widetilde{\mathbf{h}}_{\mathrm{t}}^{H}\widetilde{\mathbf{x}}(n) + v(n) + u(n), \tag{7.1}$$

where $u(n) = u_{\mathrm{L}}(n) + ju_{\mathrm{R}}(n)$ with $u_{\mathrm{L}}(n)$ and $u_{\mathrm{R}}(n)$ being the near-end signals picked up by the left and right microphones. Most often the echo path is modeled by an adaptive FIR filter, $\widetilde{\mathbf{h}}(n)$, that generates a replica of the echo. This echo estimate is then subtracted from the return channel and thereby cancellation is achieved. This may look like a simple straightforward system identification task for the adaptive filter. However, in most conversations, there are the so-called double-talk situations that make the identification much more problematic than what it might appear at a first glance. Double-talk occurs when the speech of the two talkers arrive simultaneously at the echo canceler, i.e. $x(n) \neq 0$ and $u(n) \neq 0$ (the situation with near-end talk only, $x(n) = 0$ and $u(n) \neq 0$, can be regarded as an "easy-to-detect" double-

J. Benesty et al.: A Perspective on Stereophonic Acoustic Echo Cancellation, STSP 4, pp. 71–79.
springerlink.com

talk case). In the double-talk situation, the near-end speech acts as a large level of uncorrelated noise to the adaptive algorithm. The disturbing near-end speech may cause the adaptive filter to diverge. Hence, annoying audible echo will pass through to the far-end. The usual way to alleviate this problem is to slow down or completely halt the filter adaptation when the presence of the near-end speech is detected. This is the very important role of the so-called DTD. The basic double-talk detection scheme is based on computing a detection statistic, ζ, and comparing it with a preset threshold, T.

DTDs basically operate in the same manner. Thus, the general procedure for handling double-talk is described by the following.

1. A detection statistic, ζ, is formed using available signals, e.g., $x(n)$, $d(n)$, $e(n)$, etc., and the estimated filter, $\widetilde{\mathbf{h}}(n)$.
2. The detection statistic, ζ, is compared to a preset threshold, T, and double-talk is declared if $\zeta < T$.
3. Once double-talk is declared, the detection is held for a minimum period of time T_{hold}. While the detection is held, the filter adaptation is disabled.
4. If $\zeta \geq T$ consecutively over a time T_{hold}, the filter resumes adaptation, while the comparison of ζ to T continues until $\zeta < T$ again.

The hold time T_{hold} in Step 3 and Step 4 is necessary to suppress detection dropouts due to the noisy behavior of the detection statistic. Although there are some possible variations, most of the DTDs keep this basic form and only differ in how to form the detection statistic.

An "optimum" decision variable ζ for double-talk detection will behave as follows:

(i) if $u(n) = 0$ (double-talk is not present), $\zeta \geq T$;
(ii) if $u(n) \neq 0$ (double-talk is present), $\zeta < T$;
(iii) ζ is insensitive to echo path variations.

The threshold T must be a constant, independent of the data. Moreover, it is desirable that the decisions are made without introducing any delay (or minimize the introduced delay) in the updating of the adaptive filter. The delayed decisions will otherwise affect the SAEC algorithm negatively.

A large number of double-talk detection schemes for the single-channel case have been proposed since the introduction of echo cancelers [1]. The Geigel algorithm [2] has proven successful in network echo cancelers; however, it does not always provide reliable performance when used in the acoustic situation. This is because it assumes a minimum echo path attenuation which may not be valid in the acoustic case. Other methods based on cross-correlation and coherence [3], [4], [5], [6], [7] have been studied which appear to be more appropriate for the acoustic application. Spectral comparing methods [8] and two-microphone solutions have also been proposed [9]. A DTD based on multi statistic testing in combination with modeling of the echo path by two filters is proposed in [10]. Next, we discuss some DTDs in the WL context.

7.2 DTDs Based on the Holder's Inequality

Let us consider the complex-valued vector

$$\mathbf{a} = \begin{bmatrix} a_0 \; a_1 \; \cdots \; a_{2L-1} \end{bmatrix}^T \tag{7.2}$$

of length $2L$. The ℓ_1, ℓ_2, and ℓ_∞ (maximum) norms [11] of the vector $\mathbf{a}$ are defined as, respectively,

$$\|\mathbf{a}\|_1 = \sum_{l=0}^{2L-1} |a_l|, \tag{7.3}$$

$$\begin{aligned} \|\mathbf{a}\|_2 &= \sqrt{\sum_{l=0}^{2L-1} |a_l|^2} \\ &= \sqrt{\mathbf{a}^H \mathbf{a}}, \end{aligned} \tag{7.4}$$

and

$$\|\mathbf{a}\|_\infty = \max_{0 \leq l \leq 2L-1} |a_l| \,. \tag{7.5}$$

It can be shown that [11]

$$1 \leq \frac{\|\mathbf{a}\|_1}{\|\mathbf{a}\|_2} \leq \sqrt{2L}, \tag{7.6}$$

$$1 \leq \frac{\|\mathbf{a}\|_1}{\|\mathbf{a}\|_\infty} \leq 2L, \tag{7.7}$$

$$1 \leq \frac{\|\mathbf{a}\|_2}{\|\mathbf{a}\|_\infty} \leq \sqrt{2L}. \tag{7.8}$$

These inequalities are very important since the ratios of different vector norms are lower and upper bounded by values independent of the characteristic of the vector.

Let $\mathbf{a}$ and $\mathbf{b}$ be two vectors of length $2L$, the Holder's inequality [11] states that

$$\left|\mathbf{a}^H \mathbf{b}\right| \leq \|\mathbf{a}\|_p \|\mathbf{b}\|_q \,, \quad \frac{1}{p} + \frac{1}{q} = 1. \tag{7.9}$$

In particular,

$$\left|\mathbf{a}^H \mathbf{b}\right| \leq \|\mathbf{a}\|_\infty \|\mathbf{b}\|_1 \,, \tag{7.10}$$

$$\left|\mathbf{a}^H \mathbf{b}\right| \leq \|\mathbf{a}\|_2 \|\mathbf{b}\|_2 \,. \tag{7.11}$$

In the single-talk scenario, the (complex) microphone signal is

$$d(n) = \widetilde{\mathbf{h}}_{\mathrm{t}}^{H}\widetilde{\mathbf{x}}(n) + v(n). \tag{7.12}$$

From (7.10) and (7.12), we get

$$\begin{aligned} |d(n)| &\leq \left|\widetilde{\mathbf{h}}_{\mathrm{t}}^{H}\widetilde{\mathbf{x}}(n)\right| + |v(n)| \\ &\leq \left\|\widetilde{\mathbf{h}}_{\mathrm{t}}\right\|_{\infty} \|\widetilde{\mathbf{x}}(n)\|_{1} + |v(n)|. \end{aligned} \tag{7.13}$$

Now, from (7.13), we can deduce a first detection statistic as

$$\zeta_1 = T_{\infty} \|\widetilde{\mathbf{x}}(n)\|_{1} + \sigma_v, \tag{7.14}$$

where T_{∞} is a threshold that obviously depends on $\left\|\widetilde{\mathbf{h}}_{\mathrm{t}}\right\|_{\infty}$. Consequently, if $\zeta_1 \geq |d(n)|$, we can state that there is no double-talk but if $\zeta_1 < |d(n)|$, we can declare double-talk.

Also, we can use (7.10) differently to obtain

$$\begin{aligned} |d(n)| &\leq \left|\widetilde{\mathbf{h}}_{\mathrm{t}}^{H}\widetilde{\mathbf{x}}(n)\right| + |v(n)| \\ &\leq \left\|\widetilde{\mathbf{h}}_{\mathrm{t}}\right\|_{1} \|\widetilde{\mathbf{x}}(n)\|_{\infty} + |v(n)|. \end{aligned} \tag{7.15}$$

Therefore, based on (7.15), a second detection statistic can be deduced as

$$\zeta_2 = T_1 \|\widetilde{\mathbf{x}}(n)\|_{\infty} + \sigma_v, \tag{7.16}$$

where T_1 is an approximation of $\left\|\widetilde{\mathbf{h}}_{\mathrm{t}}\right\|_{1}$. Thus, if $\zeta_2 < |d(n)|$, double-talk is declared but for $\zeta_2 \geq |d(n)|$, we have no near-end speech. This algorithm can be seen as a generalization of the Geigel algorithm [2] since the noise is taken into account. It is known that the detection statistic of the Geigel DTD is defined as

$$\zeta_{\mathrm{G}} = T_{\mathrm{G}} \|\widetilde{\mathbf{x}}(n)\|_{\infty} \tag{7.17}$$

and the double-talk is declared when $\zeta_{\mathrm{G}} < |d(n)|$. As we can see from (7.17), the existence of the system noise is not taken into account. Consequently, the Geigel DTD may perform poorly when the level of the background noise is high, interpreting this situation as double-talk.

Finally, using (7.11), we get

$$\begin{aligned} |d(n)| &\leq \left|\widetilde{\mathbf{h}}_{\mathrm{t}}^{H}\widetilde{\mathbf{x}}(n)\right| + |v(n)| \\ &\leq \left\|\widetilde{\mathbf{h}}_{\mathrm{t}}\right\|_{2} \|\widetilde{\mathbf{x}}(n)\|_{2} + |v(n)|. \end{aligned} \tag{7.18}$$

Based on (7.18), a third detection statistic can be defined as

$$\zeta_3 = T_2 \left\|\widetilde{\mathbf{x}}(n)\right\|_2 + \sigma_v, \tag{7.19}$$

where the threshold T_2 depends on $\left\|\widetilde{\mathbf{h}}_\mathrm{t}\right\|_2$. Again here ζ_3 is compared to $|d(n)|$. Condition $\zeta_3 < |d(n)|$ implies double-talk, otherwise there is no double-talk.

As we can see, all the previous developed DTDs are based on the Holder's inequality. The derived detection statistics [see (7.14), (7.16), and (7.19)] take into account the existence of the system noise, in terms of its variance. In practice, this parameter can be estimated during silences. The computational complexity of the proposed DTDs are similar to the Geigel algorithm. Regarding the computational complexity of (7.14) and (7.19), the required input signal norms $\left\|\widetilde{\mathbf{x}}(n)\right\|_1$ and $\left\|\widetilde{\mathbf{x}}(n)\right\|_2$ can be efficiently computed in a recursive way.

The main problem is how to choose the thresholds T_∞, T_1, and T_2. These parameters depend on $\left\|\widetilde{\mathbf{h}}_\mathrm{t}\right\|_\infty$, $\left\|\widetilde{\mathbf{h}}_\mathrm{t}\right\|_1$, and $\left\|\widetilde{\mathbf{h}}_\mathrm{t}\right\|_2$, respectively, which are unavailable in practice. However, let us remember that the threshold T_1 is similar to the Geigel threshold T_G, which is chosen assuming a minimum echo path attenuation. Also, we know the following inequalities [see (7.6) and (7.7)]:

$$\left\|\widetilde{\mathbf{h}}_\mathrm{t}\right\|_1 \leq \sqrt{2L} \left\|\widetilde{\mathbf{h}}_\mathrm{t}\right\|_2, \tag{7.20}$$

$$\left\|\widetilde{\mathbf{h}}_\mathrm{t}\right\|_1 \leq 2L \left\|\widetilde{\mathbf{h}}_\mathrm{t}\right\|_\infty. \tag{7.21}$$

Consequently, from (7.20) and (7.21), we get

$$T_2 \geq \frac{T_1}{\sqrt{2L}}, \tag{7.22}$$

$$T_\infty \geq \frac{T_1}{2L}. \tag{7.23}$$

Therefore, after we set the threshold $T_1 = T_\mathrm{G}$ (similar to the Geigel DTD), the other thresholds can be chosen based on (7.22) and (7.23). Here, it could be useful to know or estimate the sparseness degree of the echo path, i.e., the number of "active" or non-zero coefficients (denoted by L_s) [12], because it makes more sense to use L_s instead of $2L$ in (7.22) and (7.23) [13].

7.3 DTD Based on Cross-Correlation

In [3], the cross-correlation coefficient vector between the input signal vector and the error signal was proposed as a means for double-talk detection. A similar idea using the cross-correlation coefficient vector between the input signal vector and the microphone signal has proven more robust and reliable [4], [14]. This section will therefore focus on the cross-correlation coefficient vector between $\widetilde{\mathbf{x}}(n)$ and $d(n)$, which is defined as

$$\begin{aligned} \mathbf{c}_{\widetilde{\mathbf{x}}d} &= \frac{E\left[\widetilde{\mathbf{x}}(n)d^*(n)\right]}{\sqrt{E\left[|x(n)|^2\right] E\left[|d(n)|^2\right]}} \\ &= \frac{\mathbf{p}_{\widetilde{\mathbf{x}}d}}{\sigma_x \sigma_d} \\ &= \left[c_{\widetilde{\mathbf{x}}d,0} \; c_{\widetilde{\mathbf{x}}d,1} \cdots c_{\widetilde{\mathbf{x}}d,2L-1} \right]^T, \end{aligned} \tag{7.24}$$

where $c_{\widetilde{\mathbf{x}}d,l}$ is the cross-correlation coefficient between $\widetilde{x}_l(n)$ and $d(n)$.

The idea here is to compare

$$\begin{aligned} \zeta_{\text{cc}} &= \left\| \mathbf{c}_{\widetilde{\mathbf{x}}d} \right\|_\infty \\ &= \max_l \left| c_{\widetilde{\mathbf{x}}d,l} \right|, \; l = 0, 1, \ldots, 2L-1 \end{aligned} \tag{7.25}$$

to a threshold level T_{cc}. The decision rule will be very simple: if $\zeta_{\text{cc}} \geq T_{\text{cc}}$, then double-talk is not present; if $\zeta_{\text{cc}} < T_{\text{cc}}$, then double-talk is present.

Although the ℓ_∞ norm used in (7.25) is perhaps the most natural, other scalar metrics, e.g., ℓ_1, ℓ_2, could alternatively be used to assess the cross-correlation coefficient vectors. However, there is a fundamental problem here which is not linked to the type of metric used. The problem is that these cross-correlation coefficient vectors are not well normalized. Indeed, we can only say in general that $\zeta_{\text{cc}} \leq 1$. If $u(n) = 0$, that does not imply that $\zeta_{\text{cc}} = 1$ or any other known value. We do not know the value of ζ_{cc} in general. The amount of correlation will depend a great deal on the statistics of the signals and of the echo path. As a result, the best value of T_{cc} will vary a lot from one experiment to another. So there is no natural threshold level associated with the variable ζ_{cc} when $u(n) = 0$.

Next section presents a decision variable that exhibits better properties than the cross-correlation algorithm. This decision variable is formed by properly normalizing the cross-correlation vector between $\widetilde{\mathbf{x}}(n)$ and $d(n)$.

7.4 DTD Based on Normalized Cross-Correlation

There is a simple way to normalize the cross-correlation vector between a vector $\widetilde{\mathbf{x}}(n)$ and a scalar $d(n)$ in order to have a natural threshold level for ζ when $u(n) = 0$.

Suppose that $u(n) = 0$. In this case,

$$\sigma_d^2 = \widetilde{\mathbf{h}}_\mathrm{t}^H \mathbf{R}_{\widetilde{\mathbf{x}}} \widetilde{\mathbf{h}}_\mathrm{t} + \sigma_v^2, \tag{7.26}$$

where $\mathbf{R}_{\widetilde{\mathbf{x}}}$ is defined in Chapter 3. From $d(n) = \widetilde{\mathbf{h}}_\mathrm{t}^H \widetilde{\mathbf{x}}(n) + v(n)$, we deduce that

$$\mathbf{p}_{\widetilde{\mathbf{x}}d} = \mathbf{R}_{\widetilde{\mathbf{x}}} \widetilde{\mathbf{h}}_\mathrm{t} \tag{7.27}$$

and (7.26) can be rewritten as

$$\sigma_d^2 = \mathbf{p}_{\widetilde{\mathbf{x}}d}^H \mathbf{R}_{\widetilde{\mathbf{x}}}^{-1} \mathbf{p}_{\widetilde{\mathbf{x}}d} + \sigma_v^2. \tag{7.28}$$

In general, for $u(n) \neq 0$, we have

$$\sigma_d^2 = \mathbf{p}_{\widetilde{\mathbf{x}}d}^H \mathbf{R}_{\widetilde{\mathbf{x}}}^{-1} \mathbf{p}_{\widetilde{\mathbf{x}}d} + \sigma_v^2 + \sigma_u^2, \tag{7.29}$$

where $\sigma_u^2 = E\left[|u(n)|^2\right]$ is the variance of the near-end signal, $u(n)$. If we divide (7.28) by (7.29) and take the square root, we obtain the decision variable

$$\begin{aligned} \zeta_\mathrm{ncc} &= \sqrt{\mathbf{p}_{\widetilde{\mathbf{x}}d}^H \left(\sigma_d^2 \mathbf{R}_{\widetilde{\mathbf{x}}}\right)^{-1} \mathbf{p}_{\widetilde{\mathbf{x}}d} + \frac{\sigma_v^2}{\sigma_d^2}} \\ &= \sqrt{\mathbf{c}_{\widetilde{\mathbf{x}}d}^{\prime H} \mathbf{c}_{\widetilde{\mathbf{x}}d}^{\prime} + \frac{\sigma_v^2}{\sigma_d^2}}, \end{aligned} \tag{7.30}$$

where

$$\mathbf{c}_{\widetilde{\mathbf{x}}d}^{\prime} = \left(\sigma_d^2 \mathbf{R}_{\widetilde{\mathbf{x}}}\right)^{-1/2} \mathbf{p}_{\widetilde{\mathbf{x}}d} \tag{7.31}$$

is what we will call the normalized cross-correlation vector between $\widetilde{\mathbf{x}}(n)$ and $d(n)$.

Substituting (7.27) and (7.29) into (7.30), we show that the decision variable is

$$\zeta_\mathrm{ncc} = \frac{\sqrt{\mathbf{p}_{\widetilde{\mathbf{x}}d}^H \mathbf{R}_{\widetilde{\mathbf{x}}}^{-1} \mathbf{p}_{\widetilde{\mathbf{x}}d} + \sigma_v^2}}{\sqrt{\mathbf{p}_{\widetilde{\mathbf{x}}d}^H \mathbf{R}_{\widetilde{\mathbf{x}}}^{-1} \mathbf{p}_{\widetilde{\mathbf{x}}d} + \sigma_v^2 + \sigma_u^2}}. \tag{7.32}$$

We easily deduce from (7.32) that for $u(n) = 0$, $\zeta_\mathrm{ncc} = 1$ and for $u(n) \neq 0$, $\zeta_\mathrm{ncc} < 1$. We see that the natural value of the threshold, T_ncc, associated with ζ_ncc is equal to 1. Note also that ζ_ncc is not sensitive to changes of the echo path when $u(n) = 0$.

7.5 Performance Evaluation of DTDs

The role of the threshold T is essential to the performance of the double-talk detector. To select the value of T and to compare different DTDs objectively one could view the DTD as a classical binary detection problem. By doing so, it is possible to rely on the well-established detection theory. This approach to characterize DTDs was proposed in [5], [14].

The general characteristics of a binary detection scheme are as follows.

- *Probability of False Alarm* (P_f): probability of declaring detection when a target, in our case double-talk, is not present.
- *Probability of Detection* (P_d): probability of successful detection when a target is present.
- *Probability of Miss* ($P_\mathrm{m} = 1 - P_\mathrm{d}$): probability of detection failure when a target is present.

A well-designed DTD maximizes P_d while minimizing P_f even in a low SENR. In general, a higher P_d is achieved at the cost of a higher P_f. There should be a tradeoff in performance depending on the penalty or cost function of a false alarm [15].

One common approach to characterize different detection methods is to represent the detection characteristic P_d (or P_m) as a function of false alarm probability, P_f, under a given constraint on the SENR. This is known as a receiver operating characteristic (ROC). The P_f constraint can be interpreted as the maximum tolerable false alarm rate.

Evaluation of a DTD is carried out by estimating the performance parameters, P_d (or P_m) and P_f. A principle for this technique can be found in [14]. Though in the end, one should accompany these performance measures with a joint evaluation of the DTD and the SAEC. This is due to the fact that the response time of the DTD can seriously affect the performance of the SAEC and this is, in general, not shown in the ROC curve.

References

1. M. M. Sondhi, "An adaptive echo canceler," *Bell Syst. Techn. J.*, vol. XLVI, pp. 497–510, Mar. 1967.
2. D. L. Duttweiler, "A twelve-channel digital echo canceler," *IEEE Trans. Commun.*, vol. 26, pp. 647–653, May 1978.
3. H. Ye and B. X. Wu, "A new double-talk detection algorithm based on the orthogonality theorem," *IEEE Trans. Commun.*, vol. 39, pp. 1542–1545, Nov. 1991.
4. R. D. Wesel, "Cross-correlation vectors and double-talk control for echo cancellation," Unpublished work, 1994.
5. T. Gänsler, M. Hansson, C.-J. Ivarsson, and G. Salomonsson, "A double-talk detector based on coherence," *IEEE Trans. Commun.*, vol. 44, pp. 1421–1427, Nov. 1996.
6. J. Benesty, D. R. Morgan, and J. H. Cho, "A family of doubletalk detectors based on cross-correlation," in *Proc. IWAENC*, 1999, pp. 108–111.
7. J. Benesty, D. R. Morgan, and J. H. Cho, "An new class of doubletalk detectors based on cross-correlation," *IEEE Trans. Speech, Audio Process.*, vol. 8, pp 168–172, Mar. 2000.
8. J. Prado and E. Moulines, "Frequency-domain adaptive filtering with applications to acoustic echo cancellation," *Ann. Télécomun.*, vol. 49, pp. 414–428, 1994.

9. S. M. Kuo and Z. Pan, "An acoustic echo canceller adaptable during double-talk periods using two microphones," *Acoustics Lett.*, vol. 15, pp. 175–179, 1992.
10. K. Ochiai, T. Araseki, and T. Ogihara, "Echo canceler with two echo path models," *IEEE Trans. Commun.*, vol. COM-25, pp. 589–595, June 1977.
11. G. H. Golub and C. F. Van Loan, *Matrix Computations.* Baltimore, MD: The Johns Hopkins University Press, 1996.
12. C. Paleologu, J. Benesty, and S. Ciochină, *Sparse Adaptive Filters for Echo Cancellation.* San Rafael: Morgan & Claypool, 2010.
13. C. Paleologu, J. Benesty, T. Gaensler, and S. Ciochină, "A class of double-talk detectors based on the Holder inequality," in *Proc. IEEE ICASSP*, 2011, pp. 425–428.
14. J. H. Cho, D. R. Morgan, and J. Benesty, "An objective technique for evaluating doubletalk detectors in acoustic cancelers," *IEEE Trans. Speech, Audio Process.*, vol. 7, pp. 718–724, Nov. 1999.
15. J. Benesty, T. Gänsler, D. R. Morgan, M. M. Sondhi, and S. L. Gay, *Advances in Network and Acoustic Echo Cancellation.* Berlin, Germany: Springer-Verlag, 2001.

Chapter 8
Echo and Noise Suppression as a Binaural Noise Reduction Problem

This chapter deals with the important problem of residual-echo-plus-noise suppression. When reformulated with the WL model, this problem becomes similar to binaural noise reduction. The most useful filters for suppression are then derived.

8.1 Problem Formulation

The most important aspect of a SAEC is to identify the acoustic impulse responses with an adaptive filter and then cancel the stereo echo using the filter's output. For different reasons, this task is far to be perfect in practice [1]. Even though, in general, we can have a good amount of echo cancellation, the residual echo can be heard and, therefore, some more suppression is required with another filter.

The error signal, which is transmitted to the far-end room, is modelled as follows:

$$\begin{aligned} e(n) &= u(n) + \widehat{y}(n) + v(n) \\ &= u(n) + r(n), \end{aligned} \tag{8.1}$$

where $u(n)$ is the near-end (desired) signal, $\widehat{y}(n) = \widetilde{\mathbf{h}}^H(n-1)\widetilde{\mathbf{x}}(n)$ is the residual echo, and $r(n) = \widehat{y}(n) + v(n)$ is the residual-echo-plus-noise. In the rest, we will refer to $r(n)$ simply as noise. Our objective is then to attenuate $r(n)$ with a filter as much as possible without affecting $u(n)$. This task is equivalent to binaural noise reduction [2].

The signal model given in (8.1) can be put into a vector form if we accumulate M successive samples:

$$\boldsymbol{\varepsilon}(n) = \mathbf{u}(n) + \mathbf{r}(n), \tag{8.2}$$

J. Benesty et al.: A Perspective on Stereophonic Acoustic Echo Cancellation, STSP 4, pp. 81–94.
springerlink.com

where

$$\boldsymbol{\varepsilon}(n) = \left[e(n)\ e(n-1)\ \cdots\ e(n-M+1)\right]^T \tag{8.3}$$

is a vector of length M, and $\mathbf{u}(n)$ and $\mathbf{r}(n)$ are defined in a similar way to $\boldsymbol{\varepsilon}(n)$. Since $u(n)$ and $r(n)$ are uncorrelated, the correlation matrix (of size $M \times M$) of the error signal is

$$\begin{aligned}\mathbf{R}_{\boldsymbol{\varepsilon}} &= E\left[\boldsymbol{\varepsilon}(n)\boldsymbol{\varepsilon}^H(n)\right] \\ &= \mathbf{R}_{\mathbf{u}} + \mathbf{R}_{\mathbf{r}},\end{aligned} \tag{8.4}$$

where $\mathbf{R}_{\mathbf{u}} = E\left[\mathbf{u}(n)\mathbf{u}^H(n)\right]$ and $\mathbf{R}_{\mathbf{r}} = E\left[\mathbf{r}(n)\mathbf{r}^H(n)\right]$ are the correlation matrices of $\mathbf{u}(n)$ and $\mathbf{r}(n)$, respectively.

8.2 WL Model

By using the WL estimation theory, the estimate of $u(n)$ is obtained as [3], [4]

$$\begin{aligned}\widehat{u}(n) &= \mathbf{w}^H\boldsymbol{\varepsilon}(n) + \mathbf{w}'^H\boldsymbol{\varepsilon}^*(n) \\ &= \widetilde{\mathbf{w}}^H\widetilde{\boldsymbol{\varepsilon}}(n),\end{aligned} \tag{8.5}$$

where $\mathbf{w}$ and $\mathbf{w}'$ are two complex FIR filters of length M and

$$\widetilde{\mathbf{w}} = \begin{bmatrix}\mathbf{w} \\ \mathbf{w}'\end{bmatrix}, \tag{8.6}$$

$$\widetilde{\boldsymbol{\varepsilon}}(n) = \begin{bmatrix}\boldsymbol{\varepsilon}(n) \\ \boldsymbol{\varepsilon}^*(n)\end{bmatrix} \tag{8.7}$$

are the augmented WL filter and error vector, respectively, both of length $2M$. We can rewrite (8.5) as

$$\begin{aligned}\widehat{u}(n) &= \widetilde{\mathbf{w}}^H\left[\widetilde{\mathbf{u}}(n) + \widetilde{\mathbf{r}}(n)\right] \\ &= u_{\mathrm{f}}(n) + r_{\mathrm{rn}}(n),\end{aligned} \tag{8.8}$$

where $\widetilde{\mathbf{u}}(n)$ and $\widetilde{\mathbf{r}}(n)$ are defined in a similar way to $\widetilde{\boldsymbol{\varepsilon}}(n)$,

$$u_{\mathrm{f}}(n) = \widetilde{\mathbf{w}}^H\widetilde{\mathbf{u}}(n) \tag{8.9}$$

is a filtered version of the desired signal and its conjugate of M successive time samples, and

$$r_{\mathrm{rn}}(n) = \widetilde{\mathbf{w}}^H\widetilde{\mathbf{r}}(n) \tag{8.10}$$

is the residual noise. From (8.8), we see that $\widehat{u}(n)$ depends on the vector $\widetilde{\mathbf{u}}(n)$. However, our desired signal at time n is only $u(n)$ [and not the whole vector $\widetilde{\mathbf{u}}(n)$]; so we should decompose the vector $\widetilde{\mathbf{u}}(n)$ into two orthogonal vectors: one corresponding to the desired signal at time n and the other corresponding to the interference, i.e.,

$$\widetilde{\mathbf{u}}(n) = u(n)\boldsymbol{\rho}_{\widetilde{\mathbf{u}}u} + \widetilde{\mathbf{u}}_{\mathrm{i}}(n), \tag{8.11}$$

where

$$\boldsymbol{\rho}_{\widetilde{\mathbf{u}}u} = \frac{E\left[\widetilde{\mathbf{u}}(n)u^*(n)\right]}{E\left[|u(n)|^2\right]} \tag{8.12}$$

is the normalized [with respect to $u(n)$] correlation vector (of length $2M$) between $\widetilde{\mathbf{u}}(n)$ and $u(n)$,

$$\widetilde{\mathbf{u}}_{\mathrm{i}}(n) = \widetilde{\mathbf{u}}(n) - u(n)\boldsymbol{\rho}_{\widetilde{\mathbf{u}}u} \tag{8.13}$$

is the interference signal vector, and

$$E\left[\widetilde{\mathbf{u}}_{\mathrm{i}}(n)u^*(n)\right] = \mathbf{0}. \tag{8.14}$$

Substituting (8.11) into (8.8), we obtain

$$\begin{aligned}\widehat{u}(n) &= \widetilde{\mathbf{w}}^H\left[u(n)\boldsymbol{\rho}_{\widetilde{\mathbf{u}}u} + \widetilde{\mathbf{u}}_{\mathrm{i}}(n) + \widetilde{\mathbf{r}}(n)\right] \\ &= u_{\mathrm{fd}}(n) + u_{\mathrm{ri}}(n) + r_{\mathrm{rn}}(n),\end{aligned} \tag{8.15}$$

where

$$u_{\mathrm{fd}}(n) = u(n)\widetilde{\mathbf{w}}^H\boldsymbol{\rho}_{\widetilde{\mathbf{u}}u} \tag{8.16}$$

is the filtered desired signal and

$$u_{\mathrm{ri}}(n) = \widetilde{\mathbf{w}}^H\widetilde{\mathbf{u}}_{\mathrm{i}}(n) \tag{8.17}$$

is the residual interference. We observe that the estimate of the desired (near-end) signal at time n is the sum of three terms that are mutually uncorrelated. Therefore, the variance of $\widehat{u}(n)$ is

$$\sigma^2_{\widehat{u}} = \sigma^2_{u_{\mathrm{fd}}} + \sigma^2_{u_{\mathrm{ri}}} + \sigma^2_{r_{\mathrm{rn}}}, \tag{8.18}$$

where

$$\begin{aligned}\sigma^2_{u_{\mathrm{fd}}} &= \sigma^2_u\left|\widetilde{\mathbf{w}}^H\boldsymbol{\rho}_{\widetilde{\mathbf{u}}u}\right|^2 \\ &= \widetilde{\mathbf{w}}^H\mathbf{R}_{\widetilde{\mathbf{u}}_{\mathrm{d}}}\widetilde{\mathbf{w}}, \\ \sigma^2_{u_{\mathrm{ri}}} &= \widetilde{\mathbf{w}}^H\mathbf{R}_{\widetilde{\mathbf{u}}_{\mathrm{i}}}\widetilde{\mathbf{w}}\end{aligned} \tag{8.19}$$

$$= \widetilde{\mathbf{w}}^H \mathbf{R}_{\widetilde{\mathbf{u}}} \widetilde{\mathbf{w}} - \sigma_u^2 \left| \widetilde{\mathbf{w}}^H \boldsymbol{\rho}_{\widetilde{\mathbf{u}}u} \right|^2, \tag{8.20}$$

$$\sigma_{r_{\mathrm{rn}}}^2 = \widetilde{\mathbf{w}}^H \mathbf{R}_{\widetilde{\mathbf{r}}} \widetilde{\mathbf{w}}, \tag{8.21}$$

σ_u^2 is the variance of $u(n)$, $\mathbf{R}_{\widetilde{\mathbf{u}}_\mathrm{d}} = \sigma_u^2 \boldsymbol{\rho}_{\widetilde{\mathbf{u}}u} \boldsymbol{\rho}_{\widetilde{\mathbf{u}}u}^H$ is the correlation matrix (whose rank is equal to 1) of $\widetilde{\mathbf{u}}_\mathrm{d}(n) = u(n)\boldsymbol{\rho}_{\widetilde{\mathbf{u}}u}$, and $\mathbf{R}_{\widetilde{\mathbf{u}}_\mathrm{i}} = E\left[\widetilde{\mathbf{u}}_\mathrm{i}(n)\widetilde{\mathbf{u}}_\mathrm{i}^H(n)\right]$, $\mathbf{R}_{\widetilde{\mathbf{u}}} = E\left[\widetilde{\mathbf{u}}(n)\widetilde{\mathbf{u}}^H(n)\right]$, and $\mathbf{R}_{\widetilde{\mathbf{r}}} = E\left[\widetilde{\mathbf{r}}(n)\widetilde{\mathbf{r}}^H(n)\right]$ are the correlation matrices of $\widetilde{\mathbf{u}}_\mathrm{i}(n)$, $\widetilde{\mathbf{u}}(n)$, and $\widetilde{\mathbf{r}}(n)$, respectively.

It is clear from (8.15) that the objective of the residual-echo-plus-noise reduction problem is to find optimal filters that can minimize the effect of $u_{\mathrm{ri}}(n) + r_{\mathrm{rn}}(n)$ while preserving the desired signal, $u(n)$. But before deriving such filters, we first give some very useful performance measures for the evaluation of the time-domain binaural noise reduction problem with the WL model.

8.3 Performance Measures

How to assess suppression filters is a very important issue. In this section, we are going to define the most useful performance measures for suppression. We can divide these measures into two categories. The first category evaluates the noise reduction (or the residual-echo-plus-noise suppression) performance while the second one evaluates the desired (near-end) signal distortion. We are also going to discuss the very convenient MSE criterion in this context and show how it is related to the performance measures.

8.3.1 Noise Reduction

The input SNR is defined as

$$\mathrm{iSNR} = \frac{\sigma_u^2}{\sigma_r^2}, \tag{8.22}$$

where $\sigma_r^2 = E\left[|r(n)|^2\right]$ is the variance of the residual-echo-plus-noise.

To quantify the level of noise remaining at the output of the complex WL filter, we define the output SNR as the ratio of the variance of the filtered desired signal over the variance of the residual interference-plus-noise, i.e.,

$$\begin{aligned} \mathrm{oSNR}\left(\widetilde{\mathbf{w}}\right) &= \frac{\sigma_{u_{\mathrm{fd}}}^2}{\sigma_{u_{\mathrm{ri}}}^2 + \sigma_{r_{\mathrm{rn}}}^2} \\ &= \frac{\sigma_u^2 \left| \widetilde{\mathbf{w}}^H \boldsymbol{\rho}_{\widetilde{\mathbf{u}}u} \right|^2}{\widetilde{\mathbf{w}}^H \mathbf{R}_{\mathrm{in}} \widetilde{\mathbf{w}}} \end{aligned}$$

$$= \frac{\widetilde{\mathbf{w}}^H \mathbf{R}_{\widetilde{\mathbf{u}}_\mathrm{d}} \widetilde{\mathbf{w}}}{\widetilde{\mathbf{w}}^H \mathbf{R}_\mathrm{in} \widetilde{\mathbf{w}}}, \tag{8.23}$$

where

$$\mathbf{R}_\mathrm{in} = \mathbf{R}_{\widetilde{\mathbf{u}}_\mathrm{i}} + \mathbf{R}_{\widetilde{\mathbf{r}}} \tag{8.24}$$

is the interference-plus-noise covariance matrix. The objective of the noise reduction filter is to make the output SNR greater than the input SNR so that the quality of the noisy signal will be enhanced.

For the particular filter

$$\widetilde{\mathbf{w}} = \mathbf{i}_\mathrm{i} = \begin{bmatrix} 1 \ 0 \cdots 0 \end{bmatrix}^T \tag{8.25}$$

of length $2M$, we have

$$\mathrm{oSNR}\left(\mathbf{i}_\mathrm{i}\right) = \mathrm{iSNR}. \tag{8.26}$$

With the identity filter, $\mathbf{i}_\mathrm{i}$, the SNR cannot be improved.

For any two vectors $\widetilde{\mathbf{w}}$ and $\boldsymbol{\rho}_{\widetilde{\mathbf{u}}u}$ and a positive definite matrix $\mathbf{R}_\mathrm{in}$, we have

$$\left|\widetilde{\mathbf{w}}^H \boldsymbol{\rho}_{\widetilde{\mathbf{u}}u}\right|^2 \leq \left(\widetilde{\mathbf{w}}^H \mathbf{R}_\mathrm{in} \widetilde{\mathbf{w}}\right) \left(\boldsymbol{\rho}_{\widetilde{\mathbf{u}}u}^H \mathbf{R}_\mathrm{in}^{-1} \boldsymbol{\rho}_{\widetilde{\mathbf{u}}u}\right), \tag{8.27}$$

with equality if and only if $\widetilde{\mathbf{w}} = \varsigma \mathbf{R}_\mathrm{in}^{-1} \boldsymbol{\rho}_{\widetilde{\mathbf{u}}u}$, where $\varsigma (\neq 0)$ is an arbitrary factor. Using the previous inequality in (8.23), we deduce an upper bound for the output SNR:

$$\mathrm{oSNR}\left(\widetilde{\mathbf{w}}\right) \leq \sigma_u^2 \cdot \boldsymbol{\rho}_{\widetilde{\mathbf{u}}u}^H \mathbf{R}_\mathrm{in}^{-1} \boldsymbol{\rho}_{\widetilde{\mathbf{u}}u}, \ \forall \widetilde{\mathbf{w}} \tag{8.28}$$

and clearly

$$\mathrm{oSNR}\left(\mathbf{i}_\mathrm{i}\right) \leq \sigma_u^2 \cdot \boldsymbol{\rho}_{\widetilde{\mathbf{u}}u}^H \mathbf{R}_\mathrm{in}^{-1} \boldsymbol{\rho}_{\widetilde{\mathbf{u}}u}, \tag{8.29}$$

which implies that

$$\sigma_r^2 \cdot \boldsymbol{\rho}_{\widetilde{\mathbf{u}}u}^H \mathbf{R}_\mathrm{in}^{-1} \boldsymbol{\rho}_{\widetilde{\mathbf{u}}u} \geq 1. \tag{8.30}$$

The maximum output SNR is then

$$\mathrm{oSNR}_\mathrm{max} = \sigma_u^2 \cdot \boldsymbol{\rho}_{\widetilde{\mathbf{u}}u}^H \mathbf{R}_\mathrm{in}^{-1} \boldsymbol{\rho}_{\widetilde{\mathbf{u}}u} \tag{8.31}$$

and

$$\mathrm{oSNR}_\mathrm{max} \geq \mathrm{iSNR}. \tag{8.32}$$

The noise reduction factor quantifies the amount of noise being rejected by the filter. This quantity is defined as the ratio of the power of the noise at the microphone over the power of the interference-plus-noise remaining at the filter output, i.e.,

$$\xi_{\mathrm{nr}}\left(\widetilde{\mathbf{w}}\right) = \frac{\sigma_r^2}{\widetilde{\mathbf{w}}^H \mathbf{R}_{\mathrm{in}} \widetilde{\mathbf{w}}}. \tag{8.33}$$

The noise reduction factor is expected to be lower bounded by 1; otherwise, the filter amplifies the noise. The higher the value of the noise reduction factor, the more the noise is rejected. While the output SNR is upper bounded, the noise reduction factor is not.

8.3.2 Speech Distortion

Since the noise is reduced by the filtering operation, so is, in general, the desired speech. This speech reduction (or cancellation) implies, in general, speech distortion. The speech reduction factor, which is somewhat similar to the noise reduction factor, is defined as the ratio of the variance of the desired (near-end) signal at the microphone over the variance of the filtered desired signal, i.e.,

$$\begin{aligned}\xi_{\mathrm{sr}}\left(\widetilde{\mathbf{w}}\right) &= \frac{\sigma_u^2}{\sigma_{u_{\mathrm{fd}}}^2} \\ &= \frac{1}{\left|\widetilde{\mathbf{w}}^H \boldsymbol{\rho}_{\widetilde{\mathbf{u}}u}\right|^2}.\end{aligned} \tag{8.34}$$

A key observation is that the design of filters that do not cancel the desired signal requires the constraint

$$\widetilde{\mathbf{w}}^H \boldsymbol{\rho}_{\widetilde{\mathbf{u}}u} = 1. \tag{8.35}$$

Thus, the speech reduction factor is equal to 1 if there is no distortion and expected to be greater than 1 when distortion happens.

Another way to measure the distortion of the desired speech signal due to the filtering operation is the speech distortion index, which is defined as the MSE between the desired signal and the filtered desired signal, normalized by the variance of the desired signal, i.e.,

$$\begin{aligned}\upsilon_{\mathrm{sd}}\left(\widetilde{\mathbf{w}}\right) &= \frac{E\left\{\left|u_{\mathrm{fd}}(n) - u(n)\right|^2\right\}}{\sigma_u^2} \\ &= \left|\widetilde{\mathbf{w}}^H \boldsymbol{\rho}_{\widetilde{\mathbf{u}}u} - 1\right|^2.\end{aligned} \tag{8.36}$$

We also see from this measure that the design of filters that do not distort the desired signal requires the constraint

$$\upsilon_{\text{sd}}(\widetilde{\mathbf{w}}) = 0. \tag{8.37}$$

Therefore, the speech distortion index is equal to 0 if there is no distortion and expected to be greater than 0 when distortion occurs.

It is easy to verify that we have the following fundamental relation:

$$\frac{\text{oSNR}(\widetilde{\mathbf{w}})}{\text{iSNR}} = \frac{\xi_{\text{nr}}(\widetilde{\mathbf{w}})}{\xi_{\text{sr}}(\widetilde{\mathbf{w}})}. \tag{8.38}$$

When no distortion occurs in the desired signal, the gain in SNR coincides with the noise reduction factor. Expression (8.38) indicates the equivalence between gain/loss in SNR and distortion. In other words, a gain in SNR can be achieved only if the desired signal and/or noise are/is distorted.

8.3.3 MSE Criterion

Error criteria play a critical role in deriving optimal filters. The MSE is, by far, the most practical one.

We define the error signal between the estimated and desired signals as

$$\begin{aligned}\mathcal{E}(n) &= \widehat{u}(n) - u(n) \\ &= u_{\text{fd}}(n) + u_{\text{ri}}(n) + r_{\text{rn}}(n) - u(n),\end{aligned} \tag{8.39}$$

which can be written as the sum of two uncorrelated error signals:

$$\mathcal{E}(n) = \mathcal{E}_{\text{d}}(n) + \mathcal{E}_{\text{r}}(n), \tag{8.40}$$

where

$$\begin{aligned}\mathcal{E}_{\text{d}}(n) &= u_{\text{fd}}(n) - u(n) \\ &= \left(\widetilde{\mathbf{w}}^H \boldsymbol{\rho}_{\widetilde{\mathbf{u}}u} - 1\right) u(n)\end{aligned} \tag{8.41}$$

is the signal distortion due to the filter and

$$\begin{aligned}\mathcal{E}_{\text{r}}(n) &= u_{\text{ri}}(n) + r_{\text{rn}}(n) \\ &= \widetilde{\mathbf{w}}^H \mathbf{u}_{\text{i}}(n) + \widetilde{\mathbf{w}}^H \mathbf{r}(n)\end{aligned} \tag{8.42}$$

represents the residual interference-plus-noise.

The MSE criterion is then

$$\begin{aligned}J(\widetilde{\mathbf{w}}) &= E\left[|\mathcal{E}(n)|^2\right] \\ &= J_{\text{d}}(\widetilde{\mathbf{w}}) + J_{\text{r}}(\widetilde{\mathbf{w}}),\end{aligned} \tag{8.43}$$

where

$$\begin{aligned} J_{\mathrm{d}}(\widetilde{\mathbf{w}}) &= E\left[\left|\mathcal{E}_{\mathrm{d}}(n)\right|^2\right] \\ &= \sigma_u^2 \left|\widetilde{\mathbf{w}}^H \boldsymbol{\rho}_{\widetilde{\mathbf{u}}u} - 1\right|^2 \end{aligned} \tag{8.44}$$

and

$$\begin{aligned} J_{\mathrm{r}}(\widetilde{\mathbf{w}}) &= E\left[\left|\mathcal{E}_{\mathrm{r}}(n)\right|^2\right] \\ &= \widetilde{\mathbf{w}}^H \mathbf{R}_{\mathrm{in}} \widetilde{\mathbf{w}}. \end{aligned} \tag{8.45}$$

Two particular filters are of great interest: $\widetilde{\mathbf{w}} = \mathbf{i}_{\mathrm{i}}$ and $\widetilde{\mathbf{w}} = \mathbf{0}$. With the first one (identity filter), we have neither noise reduction nor speech distortion and with the second one (zero filter), we have maximum noise reduction and maximum speech distortion (i.e., the desired speech signal is completely nulled out). For both filters, however, it can be verified that the output SNR is equal to the input SNR. For these two particular filters, the MSEs are

$$J(\mathbf{i}_{\mathrm{i}}) = J_{\mathrm{r}}(\mathbf{i}_{\mathrm{i}}) = \sigma_r^2, \tag{8.46}$$

$$J(\mathbf{0}) = J_{\mathrm{d}}(\mathbf{0}) = \sigma_u^2. \tag{8.47}$$

As a result,

$$\mathrm{iSNR} = \frac{J(\mathbf{0})}{J(\mathbf{i}_{\mathrm{i}})}. \tag{8.48}$$

We define the normalized MSE (NMSE) with respect to $J(\mathbf{i}_{\mathrm{i}})$ as

$$\begin{aligned} J_{\mathrm{n},1}(\widetilde{\mathbf{w}}) &= \frac{J(\widetilde{\mathbf{w}})}{J(\mathbf{i}_{\mathrm{i}})} \\ &= \mathrm{iSNR} \cdot \upsilon_{\mathrm{sd}}(\widetilde{\mathbf{w}}) + \frac{1}{\xi_{\mathrm{nr}}(\widetilde{\mathbf{w}})} \\ &= \mathrm{iSNR}\left[\upsilon_{\mathrm{sd}}(\widetilde{\mathbf{w}}) + \frac{1}{\mathrm{oSNR}(\widetilde{\mathbf{w}}) \cdot \xi_{\mathrm{sr}}(\widetilde{\mathbf{w}})}\right], \end{aligned} \tag{8.49}$$

where

$$\upsilon_{\mathrm{sd}}(\widetilde{\mathbf{w}}) = \frac{J_{\mathrm{d}}(\widetilde{\mathbf{w}})}{J_{\mathrm{d}}(\mathbf{0})}, \tag{8.50}$$

$$\mathrm{iSNR} \cdot \upsilon_{\mathrm{sd}}(\widetilde{\mathbf{w}}) = \frac{J_{\mathrm{d}}(\widetilde{\mathbf{w}})}{J_{\mathrm{r}}(\mathbf{i}_{\mathrm{i}})}, \tag{8.51}$$

$$\xi_{\mathrm{nr}}(\widetilde{\mathbf{w}}) = \frac{J_{\mathrm{r}}(\mathbf{i}_{\mathrm{i}})}{J_{\mathrm{r}}(\widetilde{\mathbf{w}})}, \tag{8.52}$$

$$\mathrm{oSNR}(\widetilde{\mathbf{w}}) \cdot \xi_{\mathrm{sr}}(\widetilde{\mathbf{w}}) = \frac{J_{\mathrm{d}}(\mathbf{0})}{J_{\mathrm{r}}(\widetilde{\mathbf{w}})}. \tag{8.53}$$

This shows how this NMSE and the different MSEs are related to the performance measures.

We define the NMSE with respect to $J(\mathbf{0})$ as

$$J_{\mathrm{n},2}(\widetilde{\mathbf{w}}) = \frac{J(\widetilde{\mathbf{w}})}{J(\mathbf{0})} = \upsilon_{\mathrm{sd}}(\widetilde{\mathbf{w}}) + \frac{1}{\mathrm{oSNR}(\widetilde{\mathbf{w}}) \cdot \xi_{\mathrm{sr}}(\widetilde{\mathbf{w}})} \tag{8.54}$$

and, obviously,

$$J_{\mathrm{n},1}(\widetilde{\mathbf{w}}) = \mathrm{iSNR} \cdot J_{\mathrm{n},2}(\widetilde{\mathbf{w}}). \tag{8.55}$$

We are only interested in filters for which

$$J_{\mathrm{d}}(\mathbf{i}_{\mathrm{i}}) \leq J_{\mathrm{d}}(\widetilde{\mathbf{w}}) < J_{\mathrm{d}}(\mathbf{0}), \tag{8.56}$$

$$J_{\mathrm{r}}(\mathbf{0}) < J_{\mathrm{r}}(\widetilde{\mathbf{w}}) < J_{\mathrm{r}}(\mathbf{i}_{\mathrm{i}}). \tag{8.57}$$

From the two previous expressions, we deduce that

$$0 \leq \upsilon_{\mathrm{sd}}(\widetilde{\mathbf{w}}) < 1, \tag{8.58}$$

$$1 < \xi_{\mathrm{nr}}(\widetilde{\mathbf{w}}) < \infty. \tag{8.59}$$

It is clear that the objective of noise reduction is to find optimal filters that would either minimize $J(\widetilde{\mathbf{w}})$ or minimize $J_{\mathrm{d}}(\widetilde{\mathbf{w}})$ or $J_{\mathrm{r}}(\widetilde{\mathbf{w}})$ subject to some constraint.

8.4 Optimal Filters

In this section, we are going to derive three important filters that can help mitigate the level of the residual-echo-plus-noise.

8.4.1 Maximum Signal-to-Noise Ratio (SNR)

The maximum SNR filter, $\widetilde{\mathbf{w}}_{\max}$, is obtained by maximizing the output SNR as given in (8.23) from which, we recognize the generalized Rayleigh quotient. It is well known that this quotient is maximized with the maximum eigenvector of the matrix $\mathbf{R}_{\mathrm{in}}^{-1}\mathbf{R}_{\widetilde{\mathbf{u}}_{\mathrm{d}}}$. Let us denote by $\widetilde{\lambda}_{\max}$ the maximum eigenvalue corresponding to this maximum eigenvector. Since the rank of the mentioned matrix is equal to 1, we have

$$\begin{aligned}\widetilde{\lambda}_{\max} &= \mathrm{tr}\left(\mathbf{R}_{\mathrm{in}}^{-1}\mathbf{R}_{\widetilde{\mathbf{u}}_{\mathrm{d}}}\right) \\ &= \sigma_u^2 \cdot \boldsymbol{\rho}_{\widetilde{\mathbf{u}}u}^H \mathbf{R}_{\mathrm{in}}^{-1} \boldsymbol{\rho}_{\widetilde{\mathbf{u}}u}. \end{aligned} \tag{8.60}$$

As a result,

$$\begin{aligned}\mathrm{oSNR}\left(\widetilde{\mathbf{w}}_{\max}\right) &= \widetilde{\lambda}_{\max} \\ &= \sigma_u^2 \cdot \boldsymbol{\rho}_{\widetilde{\mathbf{u}}u}^H \mathbf{R}_{\mathrm{in}}^{-1} \boldsymbol{\rho}_{\widetilde{\mathbf{u}}u}, \end{aligned} \tag{8.61}$$

which corresponds to the maximum possible output SNR, i.e., $\mathrm{oSNR}_{\max}$.

Obviously, we also have

$$\widetilde{\mathbf{w}}_{\max} = \varsigma \mathbf{R}_{\mathrm{in}}^{-1} \boldsymbol{\rho}_{\widetilde{\mathbf{u}}u}, \tag{8.62}$$

where ς is an arbitrary non-zero scaling factor. While this factor has no effect on the output SNR, it may have on the speech distortion. In fact, the two other filters derived in the rest of this section are equivalent up to this scaling factor. These filters also try to find the respective scaling factors depending on what we optimize.

8.4.2 Wiener

The Wiener filter is easily derived by taking the gradient of the MSE, $J\left(\widetilde{\mathbf{w}}\right)$ [eq. (8.43)], with respect to $\widetilde{\mathbf{w}}$ and equating the result to zero:

$$\widetilde{\mathbf{w}}_{\mathrm{W}} = \sigma_u^2 \mathbf{R}_{\widetilde{\boldsymbol{\varepsilon}}}^{-1} \boldsymbol{\rho}_{\widetilde{\mathbf{u}}u}, \tag{8.63}$$

where $\mathbf{R}_{\widetilde{\boldsymbol{\varepsilon}}} = E\left[\widetilde{\boldsymbol{\varepsilon}}(n)\widetilde{\boldsymbol{\varepsilon}}^H(n)\right]$.

The Wiener filter can also be expressed as

$$\begin{aligned}\widetilde{\mathbf{w}}_{\mathrm{W}} &= \mathbf{R}_{\widetilde{\boldsymbol{\varepsilon}}}^{-1} \mathbf{R}_{\widetilde{\mathbf{u}}} \mathbf{i}_{\mathrm{i}} \\ &= \left(\mathbf{I}_{2M} - \mathbf{R}_{\widetilde{\boldsymbol{\varepsilon}}}^{-1} \mathbf{R}_{\widetilde{\mathbf{r}}}\right) \mathbf{i}_{\mathrm{i}}, \end{aligned} \tag{8.64}$$

where $\mathbf{I}_{2M}$ is the identity matrix of size $2M \times 2M$. The above formulation depends on the second-order statistics of the error and residual-echo-plus-noise signals. The correlation matrix $\mathbf{R}_{\widetilde{\boldsymbol{\varepsilon}}}$ can be estimated from the error signal while $\mathbf{R}_{\widetilde{\mathbf{r}}}$ can be estimated in the absence of the far-end signal.

We now propose to write the general form of the Wiener filter in another way that will make it easier to compare to other optimal filters. We can verify that

$$\mathbf{R}_{\widetilde{\boldsymbol{\varepsilon}}} = \sigma_u^2 \boldsymbol{\rho}_{\widetilde{\mathbf{u}}u} \boldsymbol{\rho}_{\widetilde{\mathbf{u}}u}^H + \mathbf{R}_{\mathrm{in}}. \tag{8.65}$$

Determining the inverse of $\mathbf{R}_{\widetilde{\varepsilon}}$ from the previous expression with the Woodbury's identity, we get

$$\mathbf{R}_{\widetilde{\varepsilon}}^{-1} = \mathbf{R}_{\text{in}}^{-1} - \frac{\mathbf{R}_{\text{in}}^{-1}\boldsymbol{\rho}_{\widetilde{\mathbf{u}}u}\boldsymbol{\rho}_{\widetilde{\mathbf{u}}u}^{H}\mathbf{R}_{\text{in}}^{-1}}{\sigma_u^{-2} + \boldsymbol{\rho}_{\widetilde{\mathbf{u}}u}^{H}\mathbf{R}_{\text{in}}^{-1}\boldsymbol{\rho}_{\widetilde{\mathbf{u}}u}}. \tag{8.66}$$

Substituting (8.66) into (8.63), leads to another interesting formulation of the Wiener filter:

$$\widetilde{\mathbf{w}}_{\text{W}} = \frac{\sigma_u^2\mathbf{R}_{\text{in}}^{-1}\boldsymbol{\rho}_{\widetilde{\mathbf{u}}u}}{1 + \sigma_u^2\boldsymbol{\rho}_{\widetilde{\mathbf{u}}u}^{H}\mathbf{R}_{\text{in}}^{-1}\boldsymbol{\rho}_{\widetilde{\mathbf{u}}u}}, \tag{8.67}$$

that we can rewrite as

$$\begin{aligned}\widetilde{\mathbf{w}}_{\text{W}} &= \frac{\sigma_u^2\mathbf{R}_{\text{in}}^{-1}\boldsymbol{\rho}_{\widetilde{\mathbf{u}}u}\boldsymbol{\rho}_{\widetilde{\mathbf{u}}u}^{H}}{1+\widetilde{\lambda}_{\max}}\mathbf{i}_{\text{i}} \\ &= \frac{\mathbf{R}_{\text{in}}^{-1}\left(\mathbf{R}_{\widetilde{\varepsilon}} - \mathbf{R}_{\text{in}}\right)}{1 + \text{tr}\left[\mathbf{R}_{\text{in}}^{-1}\left(\mathbf{R}_{\widetilde{\varepsilon}} - \mathbf{R}_{\text{in}}\right)\right]}\mathbf{i}_{\text{i}} \\ &= \frac{\mathbf{R}_{\text{in}}^{-1}\mathbf{R}_{\widetilde{\varepsilon}} - \mathbf{I}_{2M}}{1 - 2M + \text{tr}\left(\mathbf{R}_{\text{in}}^{-1}\mathbf{R}_{\widetilde{\varepsilon}}\right)}\mathbf{i}_{\text{i}}.\end{aligned} \tag{8.68}$$

From (8.68), we deduce that the output SNR is

$$\begin{aligned}\text{oSNR}\left(\widetilde{\mathbf{w}}_{\text{W}}\right) &= \widetilde{\lambda}_{\max} \\ &= \text{tr}\left(\mathbf{R}_{\text{in}}^{-1}\mathbf{R}_{\widetilde{\varepsilon}}\right) - 2M.\end{aligned} \tag{8.69}$$

We observe from (8.69) that the more the amount of noise, the smaller is the output SNR.

The speech distortion index is an explicit function of the output SNR:

$$\upsilon_{\text{sd}}\left(\widetilde{\mathbf{w}}_{\text{W}}\right) = \frac{1}{\left[1 + \text{oSNR}\left(\widetilde{\mathbf{w}}_{\text{W}}\right)\right]^2} \leq 1. \tag{8.70}$$

The higher the value of $\text{oSNR}\left(\widetilde{\mathbf{w}}_{\text{W}}\right)$, the less the desired signal is distorted.

Clearly,

$$\text{oSNR}\left(\widetilde{\mathbf{w}}_{\text{W}}\right) \geq \text{iSNR}, \tag{8.71}$$

since the Wiener filter maximizes the output SNR.

It is of interest to observe that the two filters $\widetilde{\mathbf{w}}_{\max}$ and $\widetilde{\mathbf{w}}_{\text{W}}$ are equivalent up to a scaling factor. Indeed, taking

$$\varsigma = \frac{\sigma_u^2}{1+\widetilde{\lambda}_{\max}} \tag{8.72}$$

in (8.62) (maximum SNR filter), we find (8.68) (Wiener filter).

With the Wiener filter, the noise and speech reduction factors are

$$\xi_{\mathrm{nr}}\left(\widetilde{\mathbf{w}}_{\mathrm{W}}\right) = \frac{\left(1+\widetilde{\lambda}_{\max}\right)^2}{\mathrm{iSNR}\cdot\widetilde{\lambda}_{\max}} \geq \left(1+\frac{1}{\widetilde{\lambda}_{\max}}\right)^2, \tag{8.73}$$

$$\xi_{\mathrm{sr}}\left(\widetilde{\mathbf{w}}_{\mathrm{W}}\right) = \left(1+\frac{1}{\widetilde{\lambda}_{\max}}\right)^2. \tag{8.74}$$

Finally, we give the minimum NMSEs (MNMSEs):

$$J_{\mathrm{n},1}\left(\widetilde{\mathbf{w}}_{\mathrm{W}}\right) = \frac{\mathrm{iSNR}}{1+\mathrm{oSNR}\left(\widetilde{\mathbf{w}}_{\mathrm{W}}\right)} \leq 1, \tag{8.75}$$

$$J_{\mathrm{n},2}\left(\widetilde{\mathbf{w}}_{\mathrm{W}}\right) = \frac{1}{1+\mathrm{oSNR}\left(\widetilde{\mathbf{w}}_{\mathrm{W}}\right)} \leq 1. \tag{8.76}$$

8.4.3 Minimum Variance Distortionless Response (MVDR)

The celebrated minimum variance distortionless response (MVDR) filter proposed by Capon [5], [6] can be derived in this context by minimizing the MSE of the residual interference-plus-noise, $J_{\mathrm{r}}\left(\widetilde{\mathbf{w}}\right)$, with the constraint that the desired signal is not distorted. Mathematically, this is equivalent to

$$\min_{\widetilde{\mathbf{w}}} \widetilde{\mathbf{w}}^H \mathbf{R}_{\mathrm{in}} \widetilde{\mathbf{w}} \quad \text{subject to} \quad \widetilde{\mathbf{w}}^H \boldsymbol{\rho}_{\widetilde{\mathbf{u}}u} = 1, \tag{8.77}$$

for which the solution is

$$\widetilde{\mathbf{w}}_{\mathrm{MVDR}} = \frac{\mathbf{R}_{\mathrm{in}}^{-1}\boldsymbol{\rho}_{\widetilde{\mathbf{u}}u}}{\boldsymbol{\rho}_{\widetilde{\mathbf{u}}u}^H \mathbf{R}_{\mathrm{in}}^{-1}\boldsymbol{\rho}_{\widetilde{\mathbf{u}}u}}, \tag{8.78}$$

that we can rewrite as

$$\widetilde{\mathbf{w}}_{\mathrm{MVDR}} = \frac{\mathbf{R}_{\mathrm{in}}^{-1}\mathbf{R}_{\widetilde{\boldsymbol{\varepsilon}}} - \mathbf{I}_{2M}}{\mathrm{tr}\left(\mathbf{R}_{\mathrm{in}}^{-1}\mathbf{R}_{\widetilde{\boldsymbol{\varepsilon}}}\right) - 2M}\mathbf{i}_{\mathrm{i}} = \frac{\sigma_u^2 \mathbf{R}_{\mathrm{in}}^{-1}\boldsymbol{\rho}_{\widetilde{\mathbf{u}}u}}{\widetilde{\lambda}_{\max}}. \tag{8.79}$$

Alternatively, we can express the MVDR as

$$\widetilde{\mathbf{w}}_{\mathrm{MVDR}} = \frac{\mathbf{R}_{\widetilde{\boldsymbol{\varepsilon}}}^{-1}\boldsymbol{\rho}_{\widetilde{\mathbf{u}}u}}{\boldsymbol{\rho}_{\widetilde{\mathbf{u}}u}^{H}\mathbf{R}_{\widetilde{\boldsymbol{\varepsilon}}}^{-1}\boldsymbol{\rho}_{\widetilde{\mathbf{u}}u}}. \tag{8.80}$$

The Wiener and MVDR filters are simply related as follows:

$$\widetilde{\mathbf{w}}_{\mathrm{W}} = \varsigma_0 \widetilde{\mathbf{w}}_{\mathrm{MVDR}}, \tag{8.81}$$

where

$$\begin{aligned} \varsigma_0 &= \widetilde{\mathbf{w}}_{\mathrm{W}}^{H}\boldsymbol{\rho}_{\widetilde{\mathbf{u}}u} \\ &= \frac{\widetilde{\lambda}_{\max}}{1+\widetilde{\lambda}_{\max}}. \end{aligned} \tag{8.82}$$

So, the two filters $\widetilde{\mathbf{w}}_{\mathrm{W}}$ and $\widetilde{\mathbf{w}}_{\mathrm{MVDR}}$ are equivalent up to a scaling factor. From a theoretical point of view, this scaling is not significant. But from a practical point of view it can be important. Indeed, the signals are usually nonstationary and the estimations are done frame by frame, so it is essential to have this scaling factor right from one frame to another in order to avoid large distortions. Therefore, it is recommended to use the MVDR filter rather than the Wiener filter in speech enhancement applications.

It is clear that we always have

$$\mathrm{oSNR}\left(\widetilde{\mathbf{w}}_{\mathrm{MVDR}}\right) = \mathrm{oSNR}\left(\widetilde{\mathbf{w}}_{\mathrm{W}}\right), \tag{8.83}$$

$$\upsilon_{\mathrm{sd}}\left(\widetilde{\mathbf{w}}_{\mathrm{MVDR}}\right) = 0, \tag{8.84}$$

$$\xi_{\mathrm{sr}}\left(\widetilde{\mathbf{w}}_{\mathrm{MVDR}}\right) = 1, \tag{8.85}$$

$$\xi_{\mathrm{nr}}\left(\widetilde{\mathbf{w}}_{\mathrm{MVDR}}\right) = \frac{\mathrm{oSNR}\left(\widetilde{\mathbf{w}}_{\mathrm{MVDR}}\right)}{\mathrm{iSNR}} \leq \xi_{\mathrm{nr}}\left(\widetilde{\mathbf{w}}_{\mathrm{W}}\right), \tag{8.86}$$

and

$$1 \geq J_{\mathrm{n},1}\left(\widetilde{\mathbf{w}}_{\mathrm{MVDR}}\right) = \frac{\mathrm{iSNR}}{\mathrm{oSNR}\left(\widetilde{\mathbf{w}}_{\mathrm{MVDR}}\right)} \geq J_{\mathrm{n},1}\left(\widetilde{\mathbf{w}}_{\mathrm{W}}\right), \tag{8.87}$$

$$J_{\mathrm{n},2}\left(\widetilde{\mathbf{w}}_{\mathrm{MVDR}}\right) = \frac{1}{\mathrm{oSNR}\left(\widetilde{\mathbf{w}}_{\mathrm{MVDR}}\right)} \geq J_{\mathrm{n},2}\left(\widetilde{\mathbf{w}}_{\mathrm{W}}\right). \tag{8.88}$$

References

1. J. Benesty, T. Gänsler, D. R. Morgan, M. M. Sondhi, and S. L. Gay, *Advances in Network and Acoustic Echo Cancellation.* Berlin, Germany: Springer-Verlag, 2001.
2. J. Benesty, J. Chen, and Y. Huang, "Binaural noise reduction in the time domain with a stereo setup," *IEEE Trans. Audio, Speech, Language Process.*, to appear, 2011.
3. B. Picinbono and P. Chevalier, "Widely linear estimation with complex data," *IEEE Trans. Signal Process.*, vol. 43, pp. 2030–2033, Aug. 1995.
4. D. P. Mandic and S. L. Goh, *Complex Valued Nonlinear Adaptive Filters: Noncircularity, Widely Linear and Neural Models.* Wiley, 2009.

5. J. Capon, "High resolution frequency-wavenumber spectrum analysis," *Proc. IEEE*, vol. 57, pp. 1408–1418, Aug. 1969.
6. R. T. Lacoss, "Data adaptive spectral analysis methods," *Geophysics*, vol. 36, pp. 661–675, Aug. 1971.

Chapter 9
Experimental Study

The objective of this chapter is to present by means of simulations the most important features of the adaptive algorithms described in the previous chapters. To facilitate the flow of the experiments, we will follow the structure of the previous chapters, by first analyzing NLMS-based adaptive filters, then APAs, and finally the FRLS algorithm presented in Chapter 6.

9.1 Experimental Conditions

All experiments are performed in the context of SAEC, as described in Fig. 2.1 (see Chapter 2). The acoustic impulse responses used for the far-end and near-end locations are shown in Fig. 9.1. Impulse responses in the far-end [i.e., $g_{\mathrm{L}}(n)$ and $g_{\mathrm{R}}(n)$] have 2048 coefficients, while the length of the impulse responses in the near-end [i.e., $h_{\mathrm{t,LL}}(n)$, $h_{\mathrm{t,RL}}(n)$, $h_{\mathrm{t,LR}}(n)$, and $h_{\mathrm{t,RR}}(n)$] is $L = 512$. The length of the WL adaptive filters used in the experiments is $2L = 1024$. Sample rate in all cases is 8 kHz.

Two source signals are used; a white Gaussian signal and a speech sequence. Background noise in near-end is independent white Gaussian distributed, whose level is set such that SENR = 30 dB [see (2.19) in Chapter 2]. In some experiments, an SENR = 10 dB is also evaluated. All simulations are performed in the single-talk scenario, i.e., absence of a near-end talker. In order to evaluate the tracking capabilities of the algorithms, an echo path change scenario is simulated in some experiments by shifting the impulse responses in the near-end location to the right by 12 samples.

The performance of the algorithms is evaluated in terms of two measures; (a) the normalized misalignment (in dB), computed according to (2.24), and (b) MSE averaged over 256 points for the purpose of smoothing the results.

J. Benesty et al.: A Perspective on Stereophonic Acoustic Echo Cancellation, STSP 4, pp. 95–135.
springerlink.com

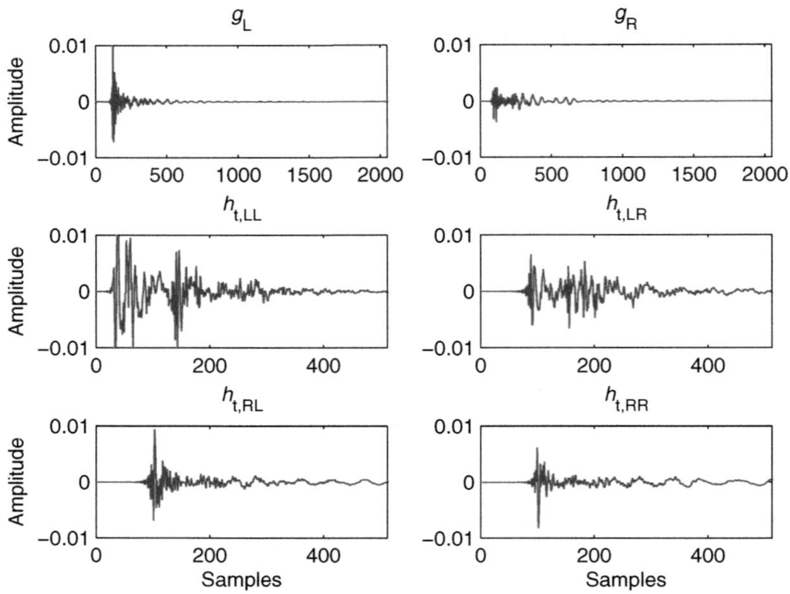

Fig. 9.1 Acoustic impulse responses used in simulations.

9.2 NLMS, VSS-NLMS, IPNLMS, and VSS-IPNLMS Algorithms

The NLMS-based algorithms (including proportionate-type algorithms and VSS versions) are typical choices for single-channel acoustic echo cancellation, due to their robustness and moderate computational complexity. However, in the multichannel case, and particularly for SAEC, there is a specific problem challenging these adaptive algorithms, namely, the strong correlation between the input (near-end loudspeaker or far-end microphone) signals $x_L(n)$ and $x_R(n)$ (see Fig. 2.1 in Chapter 2). This may result in the nonuniqueness problem (as described in Chapter 3) and, consequently, some preprocessing of these signals is, in general, necessary in order to weaken the coherence.

For the first experiment, the source signal is white Gaussian and we do not preprocess the far-end microphone signals $x_L(n)$ and $x_R(n)$. Figure 9.2 shows the misalignment of the NLMS algorithm for different values of the normalized step size ($\alpha = 1,\ 0.25,\ 0.05$) and the associated MSE curves are depicted in Fig. 9.3. The regularization parameter of the NLMS algorithm is set to $\delta = 20\sigma_x^2$, which is a practical ad-hoc choice in many echo cancellation scenario; if not specified otherwise, this value will be used in all the following experiments of this section. As we can notice from Fig. 9.2, the misalignment level is large (around -5.5 dB), no matter the value of the normalized step

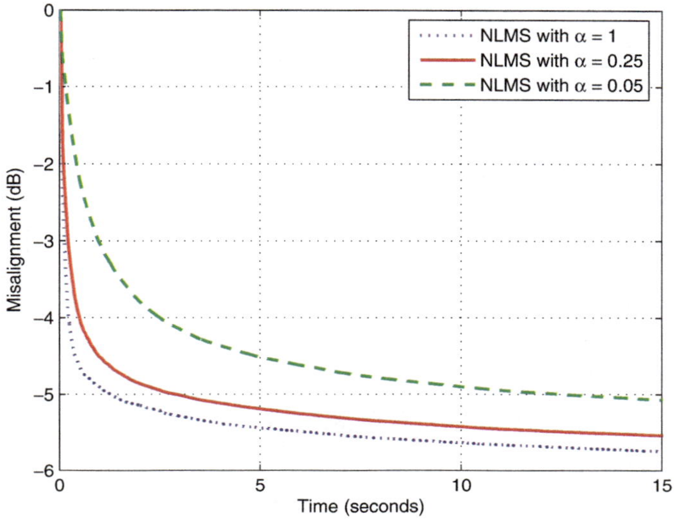

Fig. 9.2 Misalignment of the NLMS algorithm for different values of the normalized step size. The source signal is white Gaussian and is not preprocessed.

size. As expected, and seen in Fig. 9.3, the NLMS algorithm using the largest normalized step size (i.e., $\alpha = 1$) is the fastest to converge but achieves the largest MSE. However, the normalized step size $\alpha = 0.25$ better compromises between these two criteria. For this reason, this value of α will be used in all the following experiments.

As it was discussed in Chapter 3, it may be required to distort the input signals $x_{\mathrm{L}}(n)$ and $x_{\mathrm{R}}(n)$, in order to have a unique solution to the SAEC problem. Reducing the coherence between these two signals will lead to a better estimate of the true acoustic impulse responses. Of course, this distortion should be performed without affecting too much the quality of the signals and the stereo effect. A simple but efficient method uses positive and negative half-wave rectifiers on each channel respectively [2], according to (3.28) and (3.29). In this case, the amount of nonlinearity is controlled by the parameter α_{r}. In order to evaluate the influence of this approach, a second experiment is performed using different values for this parameter, i.e., $\alpha_{\mathrm{r}} = 0$ (without distortion), $\alpha_{\mathrm{r}} = 0.3$, and $\alpha_{\mathrm{r}} = 0.5$. Figure 9.4 shows the misalignments of the NLMS algorithm with the normalized step size $\alpha = 0.25$ using a white Gaussian source signal, while the corresponding MSE curves are given in Fig. 9.5. It can be noticed from Fig. 9.4 that the misalignment of the NLMS algorithm decreases when the parameter α_{r} increases. Clearly, this nonlinear distortion improves the performance in terms of the misalignment. However, according to Fig. 9.5, the MSE increases with α_{r}.

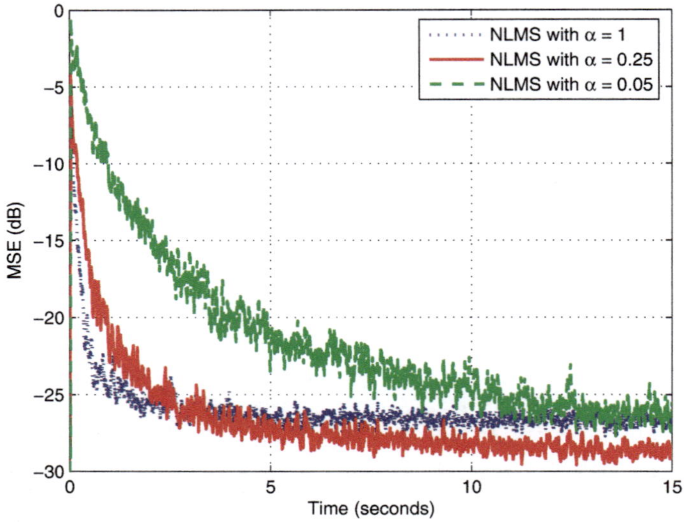

Fig. 9.3 MSE of the NLMS algorithm for different values of the normalized step size. Other conditions same as in Fig. 9.2.

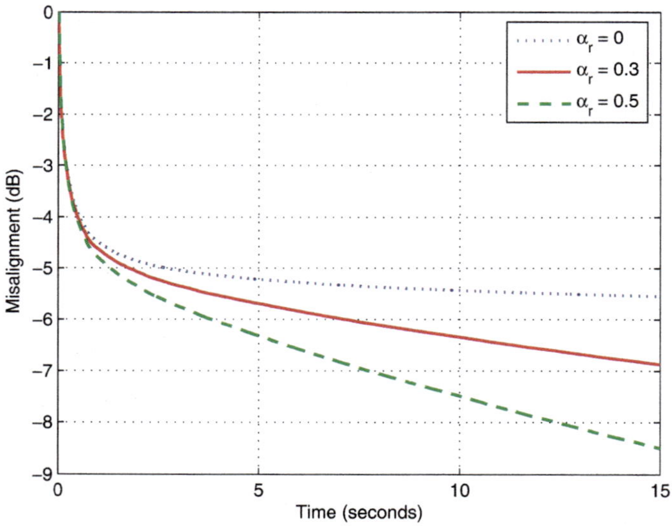

Fig. 9.4 Misalignment of the NLMS algorithm with $\alpha = 0.25$. The source signal signal is white Gaussian. Preprocessing with positive and negative half-wave rectifiers and different values of the parameter α_{r}.

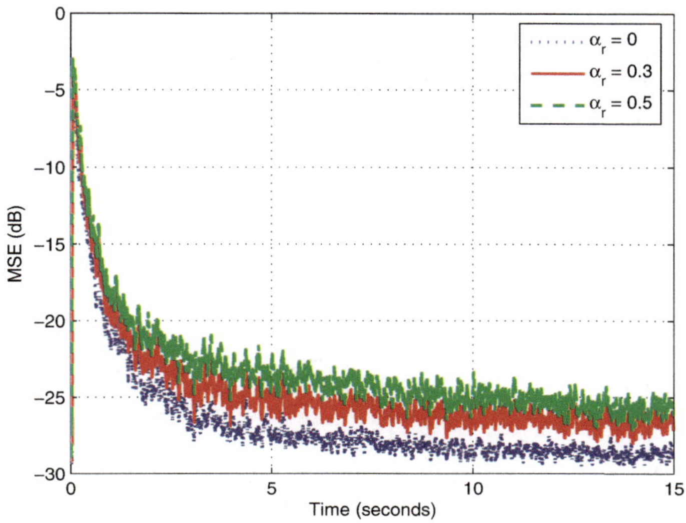

Fig. 9.5 MSE of the NLMS algorithm with $\alpha = 0.25$. Other conditions same as in Fig. 9.4.

In the context of the WL model, a new distortion was proposed in Chapter 3 [see (3.38) and (3.39)]. In this approach, the module of the complex input signal $x(n)$ is not modified, but only its phase is changed. Figure 9.6 compares the misalignment of the NLMS algorithm using positive and negative half-wave rectifiers versus the new distortion; also, the case without distortion is shown as a reference. The source is a speech sequence and the distortion parameter is set to $\alpha_{\mathrm{r}} = 0.3$. The corresponding MSE curves are depicted in Fig. 9.7. It can be noticed from Fig. 9.6 that the misalignment is greatly reduced by the new distortion. Also, as we can see in Fig. 9.7 and in the detail presented in Fig. 9.8, the new distortion leads to a better performance in terms of the MSE as compared to the positive and negative half-wave rectifiers method.

In order to justify this behavior, we depicted in Fig. 9.9 the coherence function between the two channels (estimated using the Welch method) in the context of the previous experiment. We should remember that the magnitude squared coherence between two processes is equal to 1 if and only if they are truly linearly related. According to Fig. 9.9, the new distortion leads to a weaker coherence between the channels compared to the positive and negative half-wave rectifiers. This difference is visible especially at higher frequencies; from the perceptual point of view, this is a good feature when dealing with speech signals. Figures 9.10 and 9.11 show the magnitude squared coherence for the two distortion methods, i.e., the positive and negative half-wave rectifiers and the new distortion, respectively; different values of the distortion

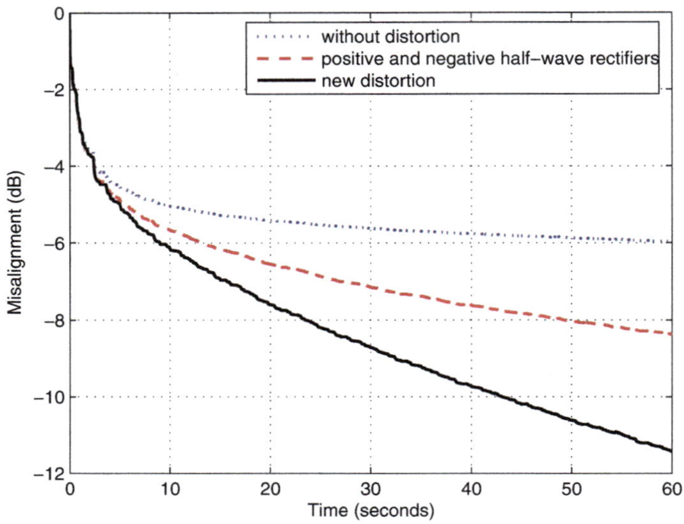

Fig. 9.6 Misalignment of the NLMS algorithm for different types of distortion with $\alpha_r = 0.3$. The source signal is a speech sequence.

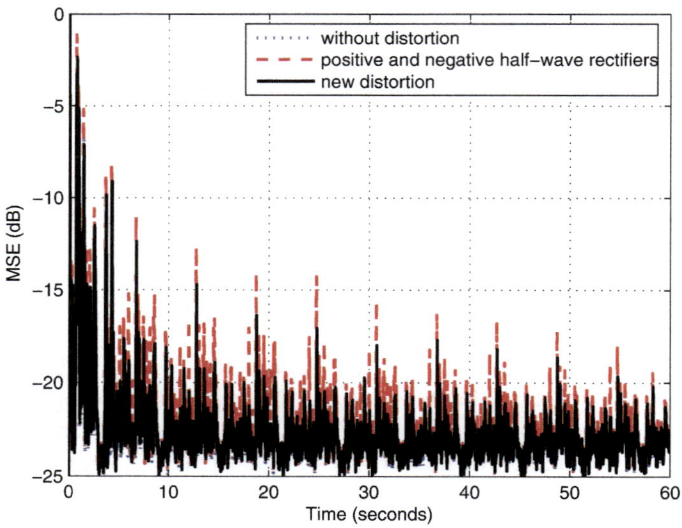

Fig. 9.7 MSE of the NLMS algorithm for different types of distortion. Other conditions same as in Fig. 9.6.

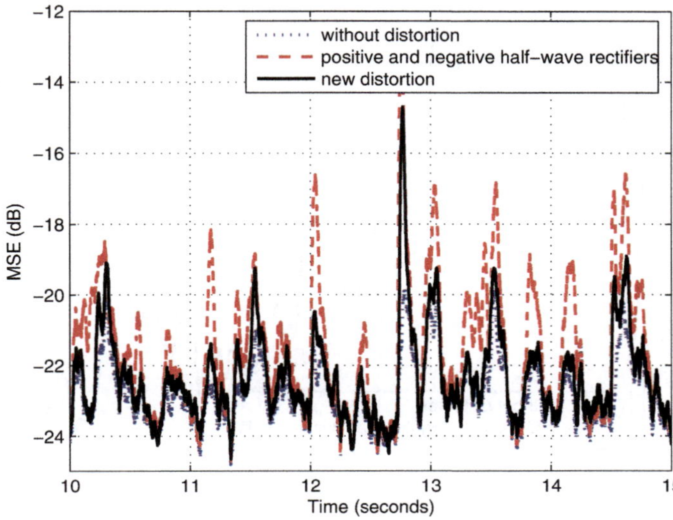

Fig. 9.8 MSE of the NLMS algorithm for different types of distortion. Detail of Fig. 9.7.

parameter α_r are used. Taking into account the previous considerations, we can even increase the value of α_r in the case of the new distortion, in order to have better performance as long as the stereo effect is not significantly affected. Subjective evaluation tests have shown that a value of $\alpha_r = 0.3$ leads to a good compromise from this point of view. Consequently, this value will be used in all the following experiments.

VSS algorithms were developed to better compromise between the convergence rate and the misadjustment, as compared to the fixed step-size algorithms. An interesting and practical VSS-NLMS algorithm [3] was presented in Section 4.6, Chapter 4. The VSS of this algorithm is evaluated according to (4.82), requiring the estimation of the system noise power, σ_v^2. In practice, this parameter could be estimated during silences; in our simulations, we assumed that its value is available. Figure 9.12 compares the misalignment of the NLMS algorithm using $\alpha = 0.25$ with the misalignment of the VSS-NLMS algorithm, while the corresponding MSE curves are given in Fig. 9.13. The source signal is white Gaussian. The resulting microphone signals are then distorted using positive and negative half-wave rectifiers with $\alpha_r = 0.3$. It can be noticed that the VSS-NLMS algorithm converges faster than the fixed step-size NLMS but achieves the same MSE level. The tracking capability of these algorithms is evaluated in Figs. 9.14 and 9.15, showing that the VSS-NLMS algorithm tracks faster than its fixed step-size counterpart.

Proportionate-type adaptive filters were found to be a very attractive choice in echo cancellation [4], [5], since they are tailored for sparse systems,

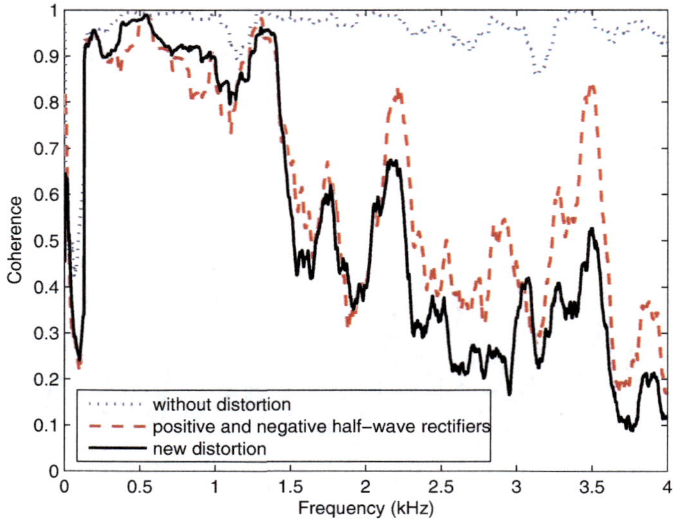

Fig. 9.9 Magnitude squared coherence function for different types of distortion with $\alpha_r = 0.3$. The source signal is a speech sequence.

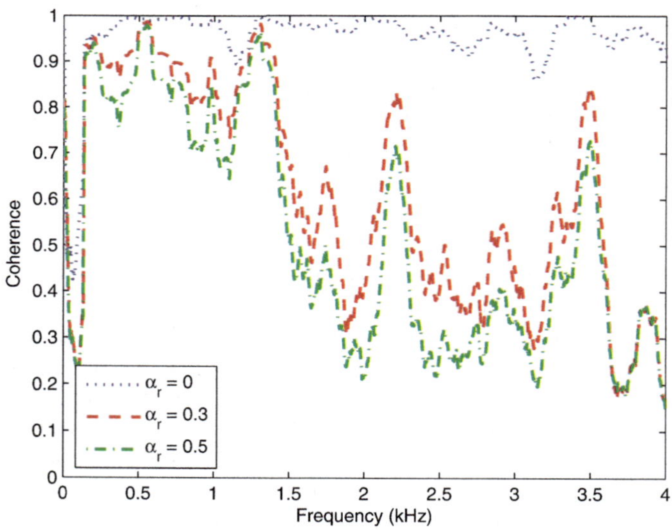

Fig. 9.10 Magnitude squared coherence function for the positive and negative half-wave rectifiers with different values of α_r. The source signal is a speech sequence.

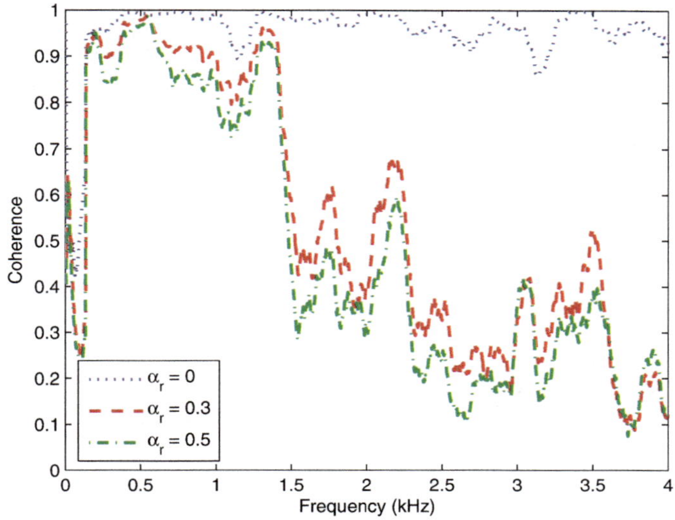

Fig. 9.11 Magnitude squared coherence function for the new distortion with different values of α_{r}. The source signal is a speech sequence.

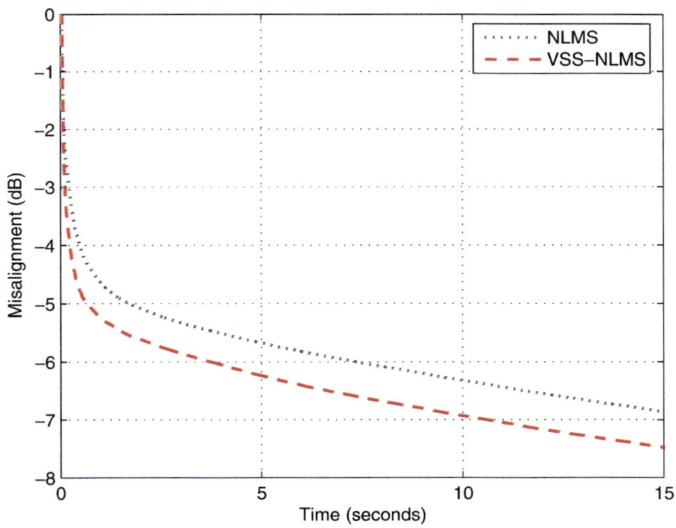

Fig. 9.12 Misalignment of the NLMS (with $\alpha = 0.25$) and VSS-NLMS algorithms. The source signal is white Gaussian. Preprocessing with positive and negative half-wave rectifiers and $\alpha_{\mathrm{r}} = 0.3$.

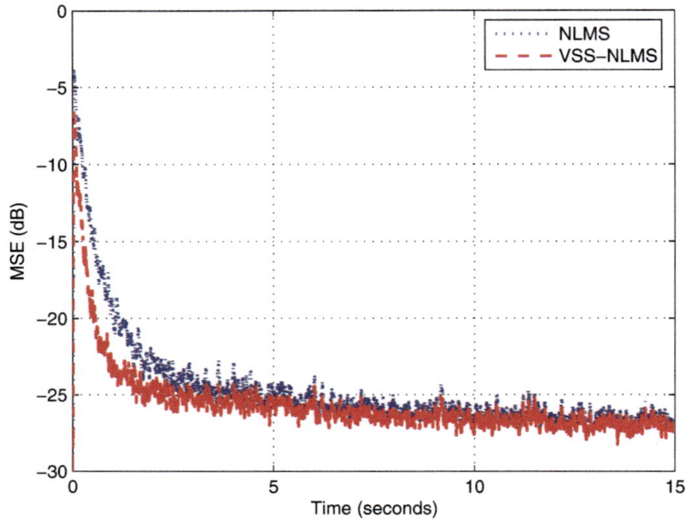

Fig. 9.13 MSE of the NLMS and VSS-NLMS algorithms. Other conditions same as in Fig. 9.12.

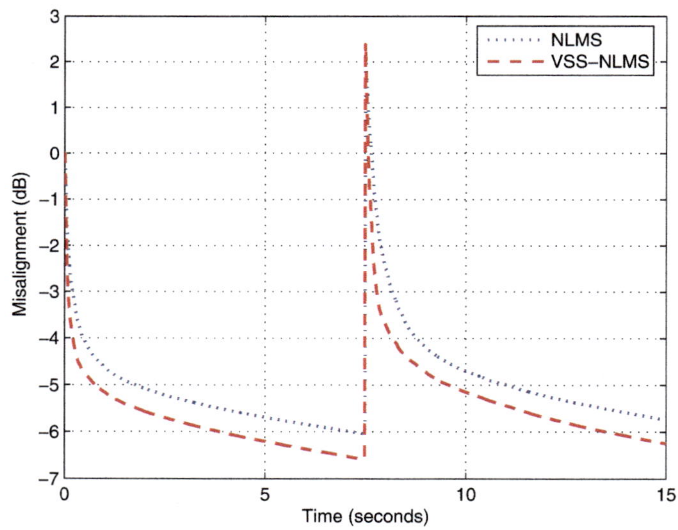

Fig. 9.14 Misalignment of the NLMS and VSS-NLMS algorithms in a tracking situation. Other conditions same as in Fig. 9.12.

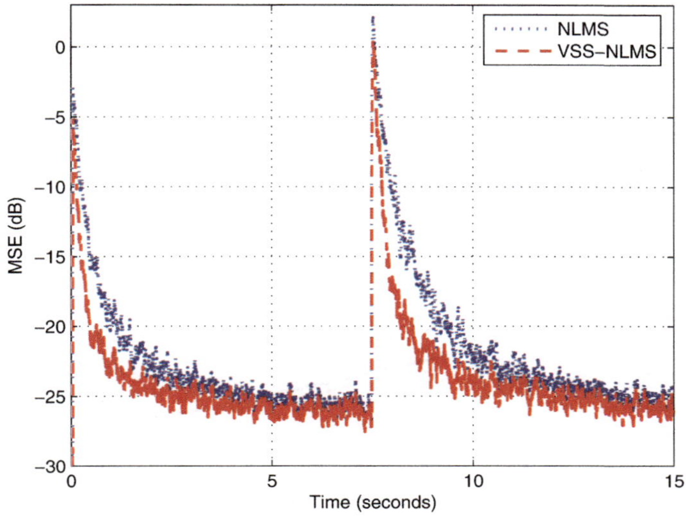

Fig. 9.15 MSE of the NLMS and VSS-NLMS algorithms in a tracking situation. Other conditions same as in Fig. 9.12.

which is the case for many echo path examples. Among many proportionate-type NLMS algorithms, the IPNLMS [6] is one of the most interesting choices, mainly due to its robustness to the sparseness degree of the echo path. The proportionate "amount" of the IPNLMS algorithm is controlled by the parameter κ ($-1 \leq \kappa < 1$) (see Section 4.7). Figure 9.16 shows the misalignment of the IPNLMS algorithm using different values of the parameter κ; the NLMS algorithm is also plotted as a reference. The corresponding MSE curves are provided in Fig. 9.17. The source signal is white Gaussian, the normalized step size for all the algorithms is $\alpha = 0.25$, and the regularization parameter of the IPNLMS is $\delta = 20\sigma_x^2/(2L)$. The far-end microphone signals are distorted using positive and negative half-wave rectifiers with $\alpha_{\rm r} = 0.3$. Figure 9.16 justifies the recommended choices for the proportionate amount, i.e., $\kappa = 0$ or -0.5 [6]. According to Fig. 9.17, all the algorithms perform very similarly in terms of the MSE. However, in the following experiments involving the IPNLMS algorithm, we will use $\kappa = 0$ since it is a more proper choice in terms of the robustness to the sparseness degree of the echo paths.

Figure 9.18 compares the misalignment of the IPNLMS algorithm using positive and negative half-wave rectifiers versus the new distortion; also, the case without distortion is shown as a reference. The input source is a speech sequence and the distortion parameter is set to $\alpha_{\rm r} = 0.3$. The corresponding MSE curves are depicted in Fig. 9.19. It can be noticed from Fig. 9.18 that the misalignment is greatly improved by the new distortion. Also, as we can see

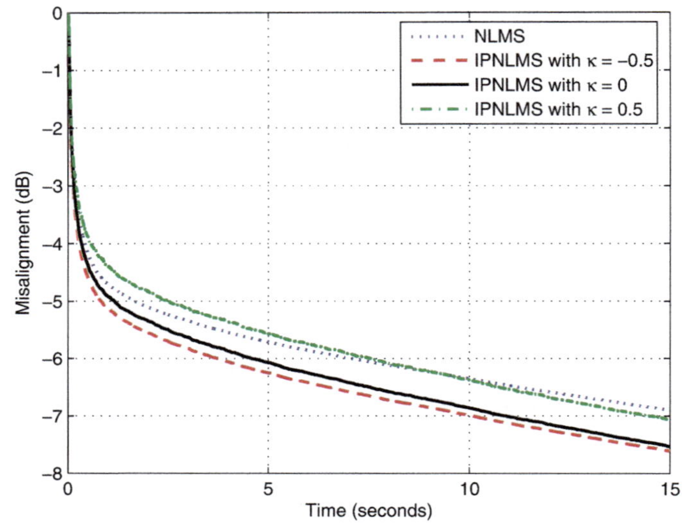

Fig. 9.16 Misalignment of the NLMS and IPNLMS algorithms. The source signal is white Gaussian. Preprocessing with positive and negative half-wave rectifiers and $\alpha_r = 0.3$.

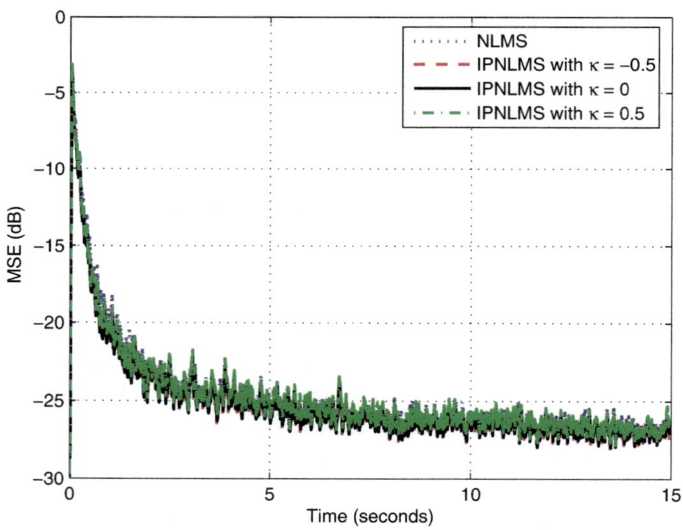

Fig. 9.17 MSE of the NLMS and IPNLMS algorithms. Other conditions same as in Fig. 9.16.

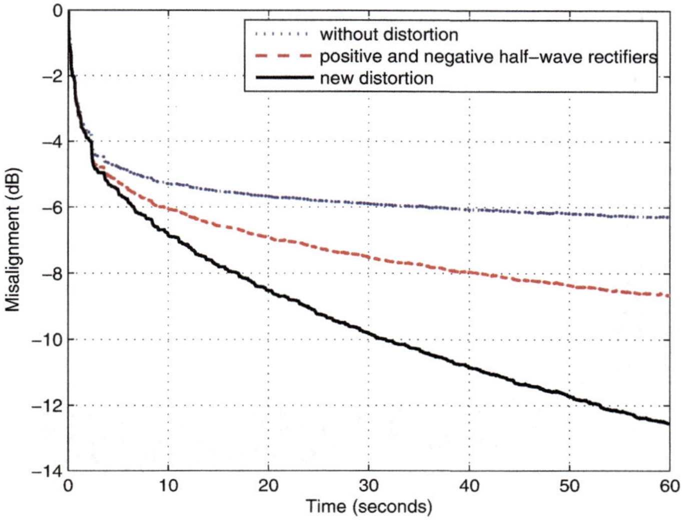

Fig. 9.18 Misalignment of the IPNLMS algorithm for different types of distortion with $\alpha_r = 0.3$. The source signal is a speech sequence.

in Fig. 9.19 and in the detail presented in Fig. 9.20, the new distortion leads to a better performance in terms of the MSE as compared to the positive and negative half-wave rectifiers.

The tracking capability of the IPNLMS algorithm is evaluated in Figs. 9.21 (for the misalignment) and 9.22 (for the MSE), as compared to the NLMS algorithm. The input source is a speech sequence and the new distortion is used with $\alpha_r = 0.3$. As we can notice, the IPNLMS tracks faster than the NLMS.

Following a similar idea as in the case of the VSS-NLMS algorithm, a VSS-IPNLMS was presented in Section 4.9, Chapter 4. The step-size of this algorithm is evaluated according to (4.106). Figure 9.23 compares the misalignment of the IPNLMS algorithm using $\alpha = 0.25$ with the misalignment of the VSS-IPNLMS algorithm, while the corresponding MSE curves are given in Fig. 9.24. The source signal is white Gaussian. The far-end microphone signals are distorted using positive and negative half-wave rectifiers with $\alpha_r = 0.3$. It can be noticed that the VSS-IPNLMS algorithm converges faster than the fixed step-size IPNLMS but achieves the same MSE level. The tracking capability of these algorithms is evaluated in Figs. 9.25 and 9.26, showing that the VSS-IPNLMS algorithm tracks faster than its fixed step-size counterpart.

Regularization is a very important issue in adaptive filtering. It is known that its importance becomes more apparent for lower values of the SENR. Based on these considerations, optimal regularization parameters for both

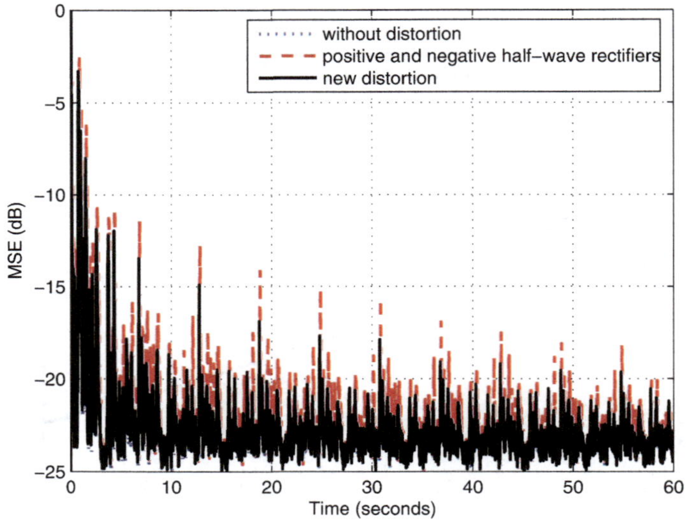

Fig. 9.19 MSE of the IPNLMS algorithm for different types of distortion. Other conditions same as in Fig. 9.18.

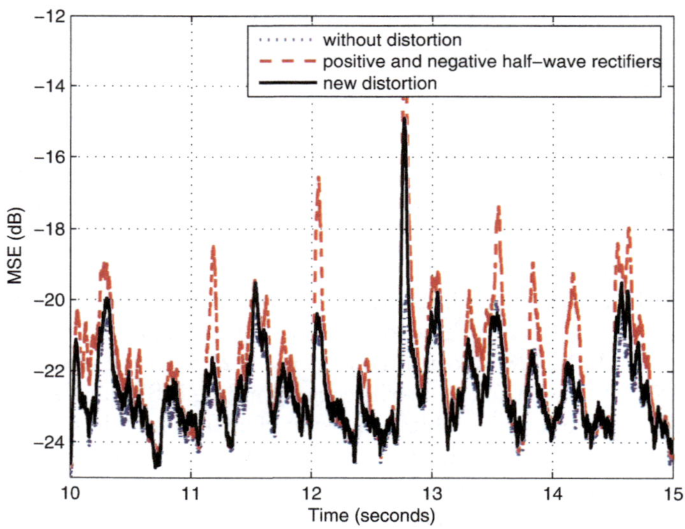

Fig. 9.20 MSE of the IPNLMS algorithm for different types of distortion. Detail of Fig. 9.19.

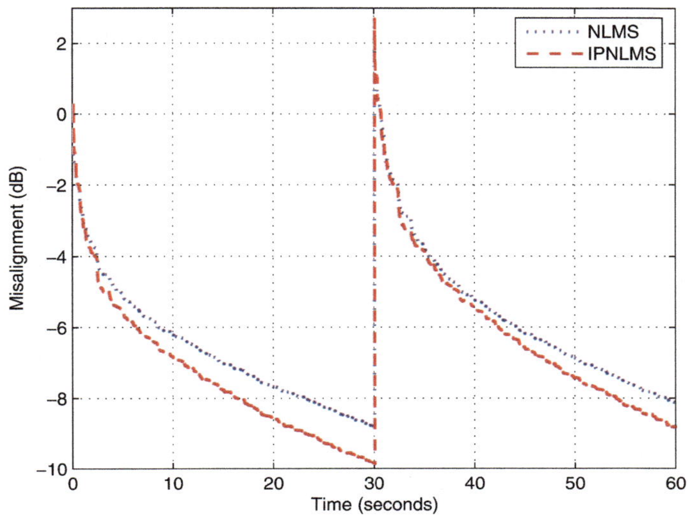

Fig. 9.21 Misalignment of the NLMS and IPNLMS algorithms in a tracking situation. The source signal is a speech sequence and the new distortion is used with $\alpha_r = 0.3$.

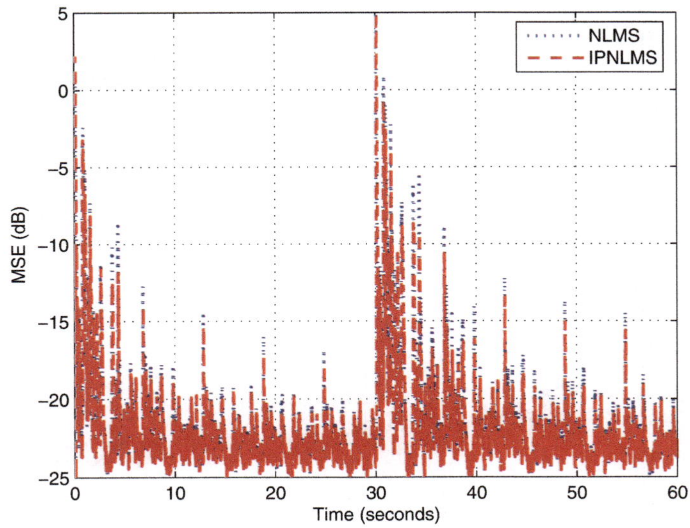

Fig. 9.22 MSE of the NLMS and IPNLMS algorithms in a tracking situation. Other conditions same as in Fig. 9.21.

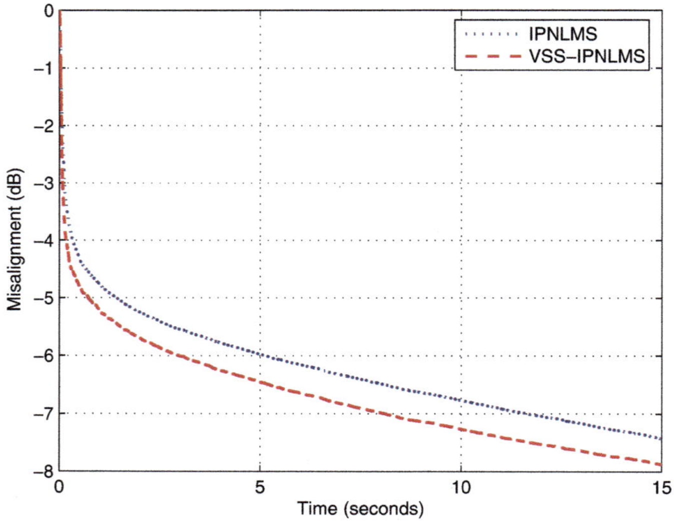

Fig. 9.23 Misalignment of the IPNLMS (with $\alpha = 0.25$) and VSS-IPNLMS algorithms. The source signal is white Gaussian. Preprocessing with positive and negative half-wave rectifiers and $\alpha_{\mathrm{r}} = 0.3$.

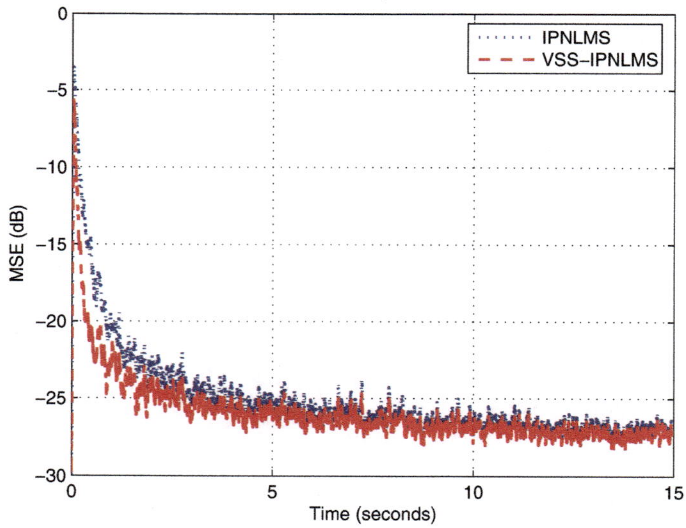

Fig. 9.24 MSE of the IPNLMS and VSS-IPNLMS algorithms. Other conditions same as in Fig. 9.23.

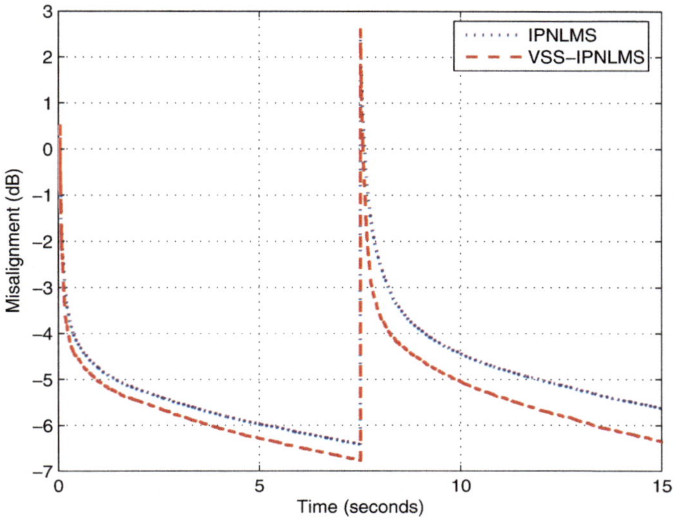

Fig. 9.25 Misalignment of the IPNLMS and VSS-IPNLMS algorithms in a tracking situation. Other conditions same as in Fig. 9.23.

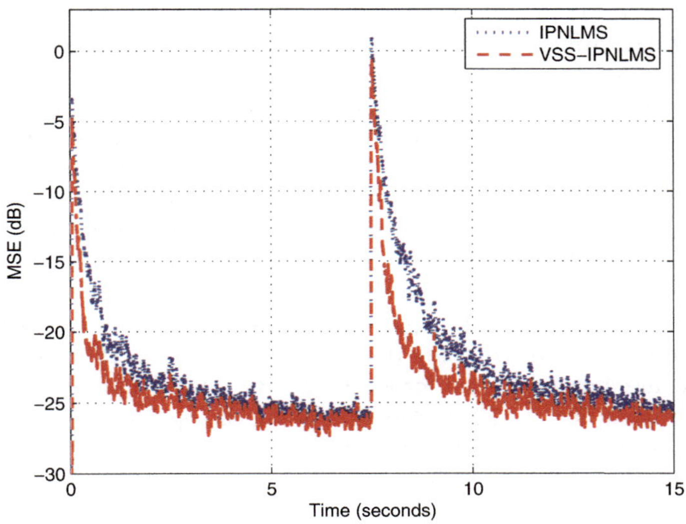

Fig. 9.26 MSE of the IPNLMS and VSS-IPNLMS algorithms in a tracking situation. Other conditions same as in Fig. 9.23.

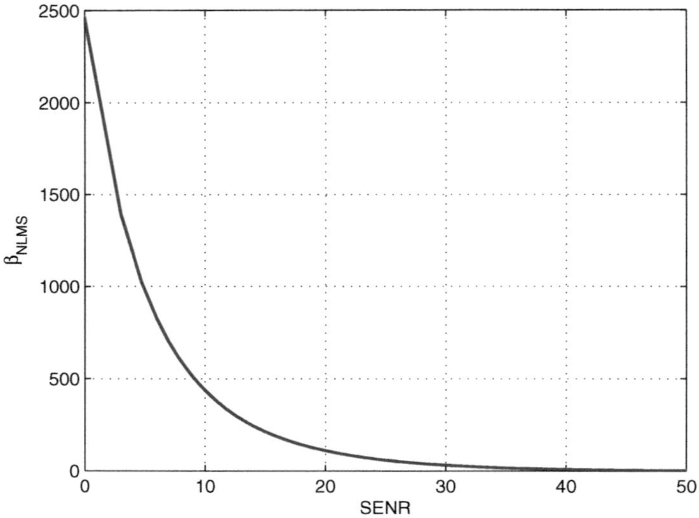

Fig. 9.27 Evolution of the normalized regularization parameter, β_{NLMS}, as a function of the SENR with $2L = 1024$. The SENR varies from 0 to 50 dB.

NLMS and IPNLMS algorithms were derived in Chapter 4. The optimal normalized regularization parameter of the NLMS algorithm, denoted by β_{NLMS}, is given in (4.65). As we can see, it depends on the SENR and the length of the adaptive filter ($2L$). In Fig. 9.27, the normalized regularization parameter β_{NLMS} is plotted for $2L = 1024$ for different values of the SENR (between 0 and 50 dB). As expected, the importance of β_{NLMS} becomes more apparent for low SENRs. Also, as it can be noticed from the detailed figure presented in Fig. 9.28, the usual "ad-hoc" choice $\beta_{\mathrm{NLMS}} = 20$ corresponds to a value of the SENR close to 30 dB, which is also an usual choice in many simulation scenarios related to echo cancellation. Consequently, the performance of the NLMS algorithm with β_{NLMS} is very similar to the case when the classical ad-hoc normalized regularization $\beta = 20$ is used. However, the difference becomes more apparent for lower SENR values. Figure 9.29 compares the misalignment of the NLMS algorithm using the optimal β_{NLMS} with the ad-hoc choice $\beta = 20$, when the SENR is set to 10 dB. The corresponding MSE curves are provided in Fig. 9.30. The source signal is speech and the new distortion is used with $\alpha_{\mathrm{r}} = 0.3$. According to these results, it is clear that the NLMS algorithm using the optimal regularization outperforms by far the classical regularization, in terms of both the misalignment and MSE.

The optimal normalized regularization parameter of the IPNLMS algorithm, denoted by β_{IPNLMS}, is given in (4.103). Also, its value depends on the SENR and the length of the adaptive filter ($2L$) but it does not depend on the proportionate parameter κ. The previous experiment is repeated in the

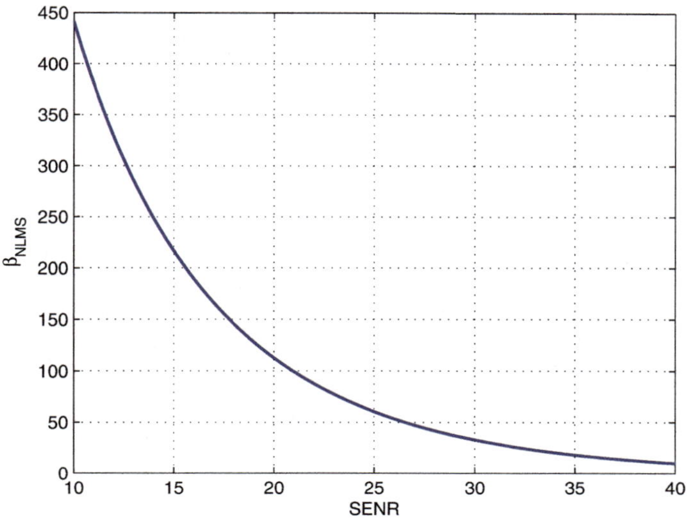

Fig. 9.28 Evolution of the normalized regularization parameter, β_{NLMS}, as a function of the SENR with $2L = 1024$. The SENR varies from 10 to 40 dB.

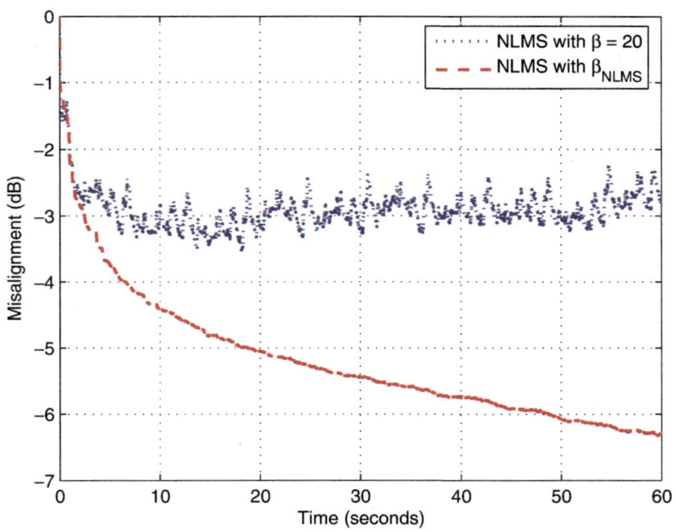

Fig. 9.29 Misalignment of the NLMS algorithm using $\beta = 20$ and β_{NLMS}. SENR = 10 dB, the source signal is speech, and the new distortion is used with $\alpha_{\mathrm{r}} = 0.3$.

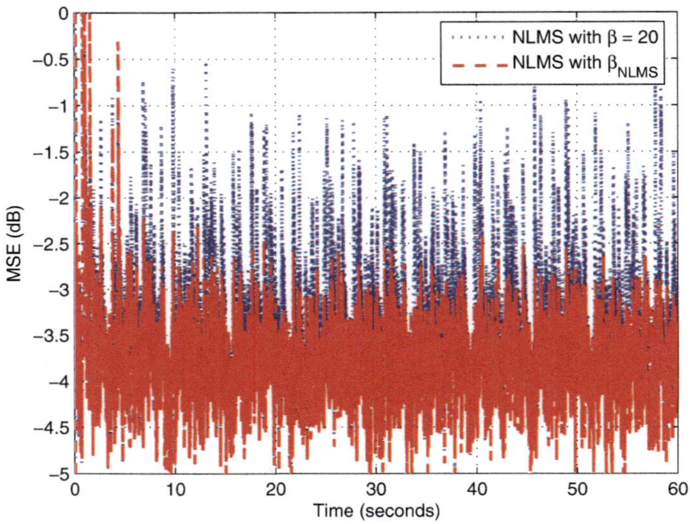

Fig. 9.30 MSE of the NLMS algorithm using $\beta = 20$ and β_{NLMS}. Other conditions same as in Fig. 9.29.

case of the IPNLMS algorithm, using the same SENR = 10 dB. The results are presented in Figs. 9.31 (for the misalignment) and 9.32 (for the MSE). The conclusion is basically the same, i.e., the IPNLMS algorithm using the optimal regularization outperforms the classical one.

Through this section, we have discussed the most important NLMS-based algorithms presented in Chapter 4. However, as a common limitation, the convergence of these algorithms is quite slow and may not be satisfactory in practical SAEC scenarios. We outline this aspect in Fig. 9.33, where the misalignment of the NLMS algorithm is plotted for a longer simulation time. The source signal is a speech sequence. It can be noticed that even when we use the new distortion, the convergence is quite slow. Taking this aspect into consideration, there is a need for faster convergence algorithms like APA or FRLS, which will be analyzed in the next two sections.

9.3 APA, VSS-APA, IPAPA, and MIPAPA

The APA [7] was derived as a generalization of the NLMS algorithm, in the sense that each tap weight vector update of the NLMS is viewed as a one dimensional affine projection, while in the APA the projections are made in multiple dimensions. When the projection dimension increases, the conver-

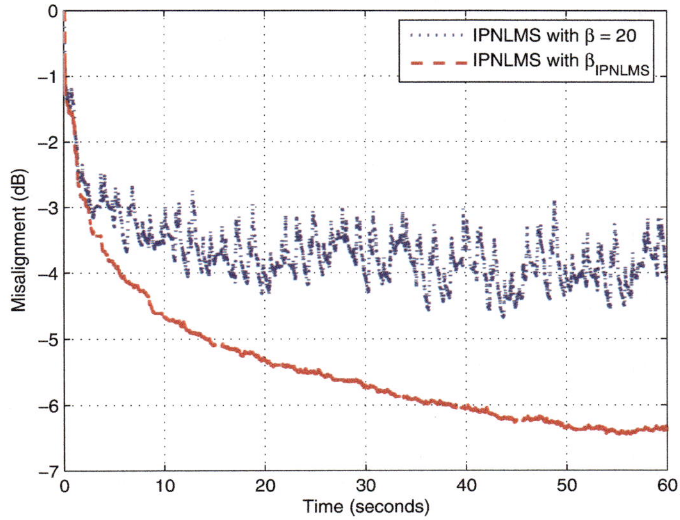

Fig. 9.31 Misalignment of the IPNLMS algorithm using $\beta = 20$ and β_{IPNLMS}. SENR = 10 dB, the source signal is speech, and the new distortion is used with $\alpha_{\mathrm{r}} = 0.3$.

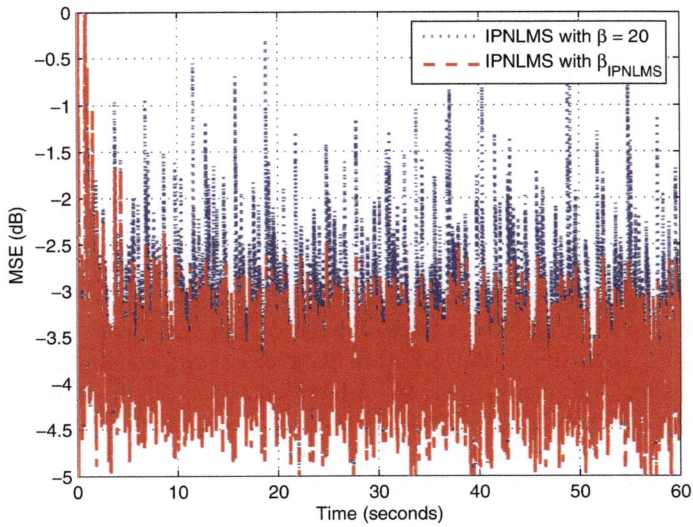

Fig. 9.32 MSE of the IPNLMS algorithm using $\beta = 20$ and β_{IPNLMS}. Other conditions same as in Fig. 9.31.

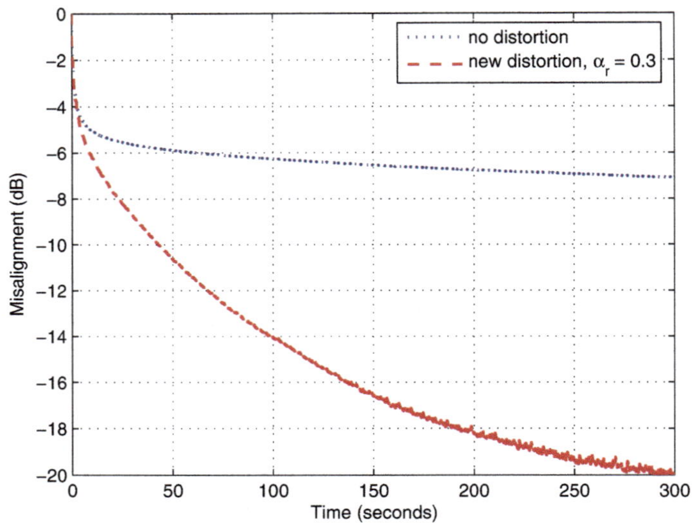

Fig. 9.33 Misalignment of the NLMS algorithm. The source signal is speech and the new distortion is used with $\alpha_{\mathrm{r}} = 0.3$.

gence rate of the tap weight vector also increases; of course, this also leads to an increased computational complexity. Nevertheless, the main advantage of the APA over the NLMS algorithm consists of a superior convergence rate especially for correlated inputs. For this reason, the APA and different versions of it were found to be very attractive choices for echo cancellation, where long filters and highly-correlated signals (like speech) are involved. Consequently, it is also expected that the APA will outperform the NLMS in the context of SAEC.

As we discussed in the previous section, the correlation between the input signals $x_{\mathrm{L}}(n)$ and $x_{\mathrm{R}}(n)$ limits the performance of the adaptive filters. The first experiment evaluates the performance of the APA without using any preprocessing (i.e., distortion). Figure 9.34 compares the misalignment of the APA using different projection orders (i.e., $P = 2,\ 8$, or 16) with the misalignment of the NLMS algorithm (which is equivalent to the APA with $P = 1$). The corresponding MSE curves are plotted in Fig. 9.35. The source signal is white Gaussian. The normalized step size for all the algorithms is set to $\alpha = 0.25$ and the regularization parameter is $\delta = 20\sigma_x^2$. If not specified otherwise, these values will be used in all the following experiments for this section. As expected, the convergence rate of the APA increases when the value of the projection order increases. However, for $P > 8$ this difference is not significant. Besides, as we can notice from Fig. 9.35, the MSE of the APA also increases with the projection order. Overall, we cannot see

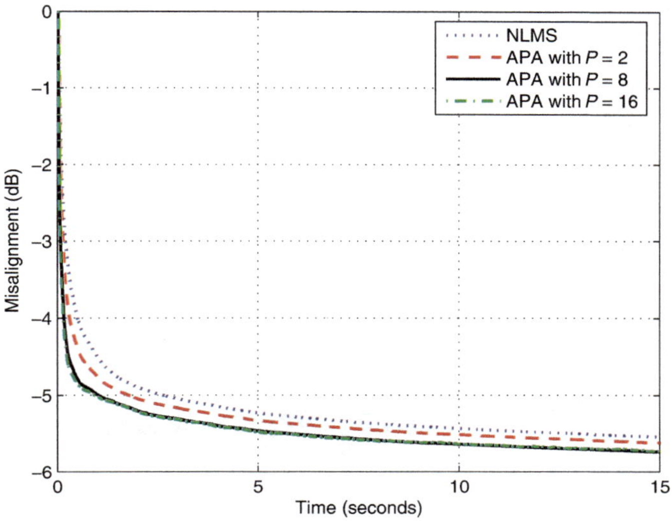

Fig. 9.34 Misalignment of the NLMS algorithm and the APA using different values of the projection order. The source signal is white Gaussian and is not preprocessed.

a significant improvement over the NLMS algorithm without preprocessing the input signal.

For this reason, the previous experiment is repeated but using positive and negative half-wave rectifiers with $\alpha_{\mathrm{r}} = 0.3$ to distort the far-end microphone signals. The results are shown in Figs. 9.36 (for the misalignment) and 9.37 (for the MSE). It can be noticed that the distortion improves the misalignment of the APA; also, the performance gain is more apparent as compared to the NLMS algorithm. This experiment also supports the idea that the projection order should not be increased too much; a value of $P = 8$ seems to offer a proper compromise between the performance and complexity. Consequently, this value of the projection order will be used in all the following experiments involving APAs.

Similar to the case of the NLMS algorithm, the distortion amount (in terms of the value of α_{r}) influences the performance of the APA. The next experiment supports this aspect, showing the performance of the APA with $P = 8$ for different values of the distortion parameter, i.e., $\alpha_{\mathrm{r}} = 0$ (without distortion), $\alpha_{\mathrm{r}} = 0.3$, and $\alpha_{\mathrm{r}} = 0.5$. Figure 9.38 shows the misalignment of the APA, while the corresponding MSE curves are given in Fig. 9.39. The source signal is white Gaussian. It can be noticed from Fig. 9.38 that the misalignment of the APA decreases when the parameter α_{r} increases. However, according to Fig. 9.39, the MSE increases with α_{r}, which indicates that a compromise should be made when choosing the value of the distortion

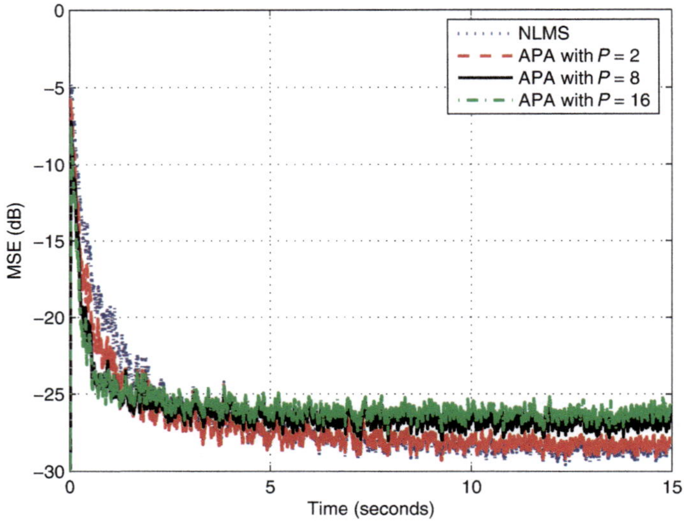

Fig. 9.35 MSE of the NLMS algorithm and the APA using different values of the projection order. Other conditions same as in Fig. 9.34.

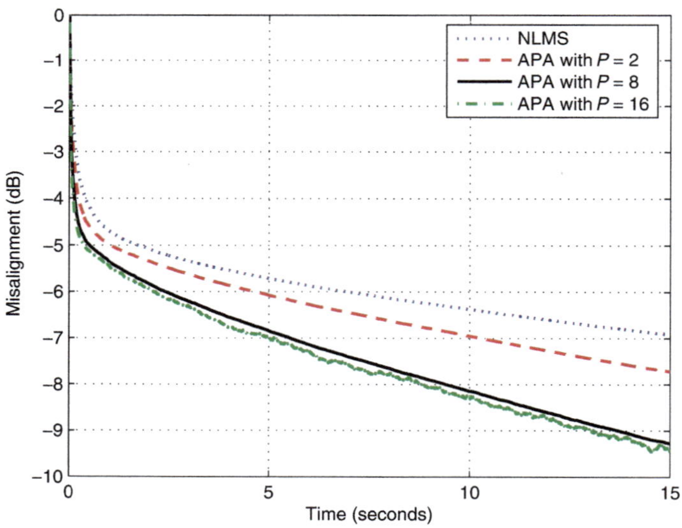

Fig. 9.36 Misalignment of the NLMS algorithm and the APA using different values of the projection order. The source signal is white Gaussian. Preprocessing with positive and negative half-wave rectifiers and $\alpha_r = 0.3$.

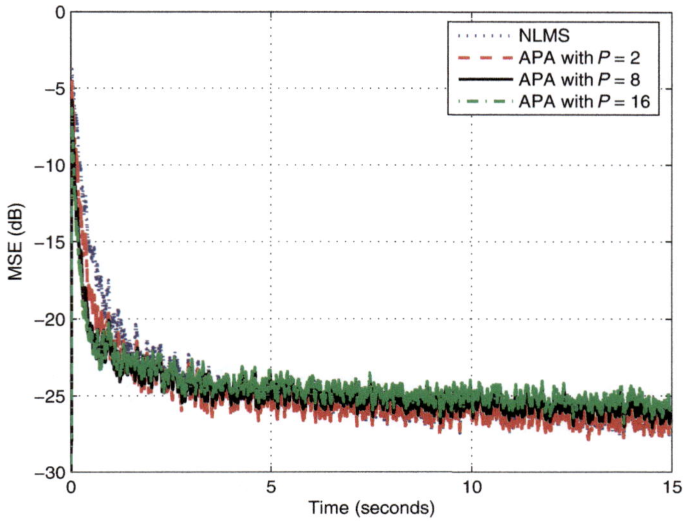

Fig. 9.37 MSE of the NLMS algorithm and the APA using different values of the projection order. Other conditions same as in Fig. 9.36.

parameter. In order to satisfy this issue, the value of $\alpha_\mathrm{r} = 0.3$ will be used in all the following experiments for this section.

Next, we evaluate the impact of the new distortion proposed in Chapter 3 [see (3.38) and (3.39)]. Figure 9.40 compares the misalignment of the APA (with $P = 8$) using positive and negative half-wave rectifiers versus the new distortion; also, the case without distortion is plotted as a reference. The input source is a speech sequence and the distortion parameter is set to $\alpha_\mathrm{r} = 0.3$. The corresponding MSE curves are depicted in Fig. 9.41. It can be noticed from Fig. 9.40 that the APA converges faster with the new distortion. Also, as we can see in Fig. 9.41 and in the detail presented in Fig. 9.42, the new distortion leads to a slightly better performance in terms of the MSE as compared to the positive and negative half-wave rectifiers.

Finally, the performance of the APA with $P = 8$ is evaluated in a tracking situation, as compared to the NLMS algorithm. The source signal is a speech sequence and the new distortion is used with $\alpha_\mathrm{r} = 0.3$. The results are provided in Figs. 9.43 (for the misalignment) and 9.44 (for the MSE). According to these plots, the APA clearly outperforms the NLMS algorithm.

Similar to the case of the VSS-NLMS algorithms, the VSS-APAs were developed to better compromise between the convergence rate and the misadjustment, as compared to the fixed step-size APAs. Such a VSS-APA [8] was presented in Section 5.4, Chapter 5. The nice feature of this algorithm is that it does not require any information about the system noise power; in

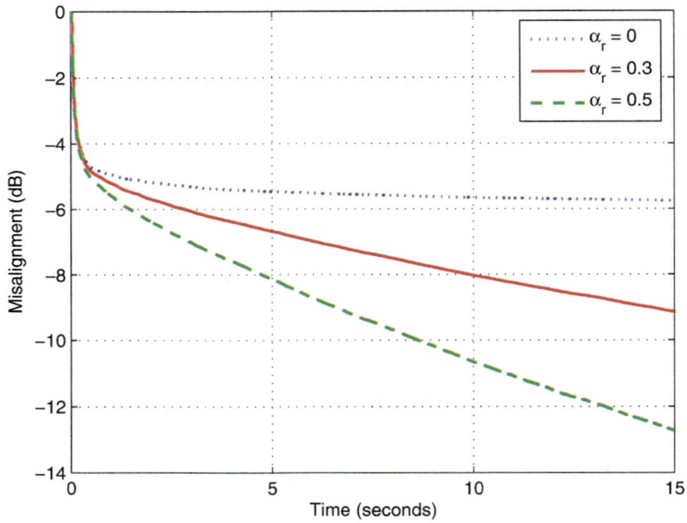

Fig. 9.38 Misalignment of the APA with $P = 8$. The source signal is white Gaussian. Preprocessing with positive and negative half-wave rectifiers and different values of the parameter α_{r}.

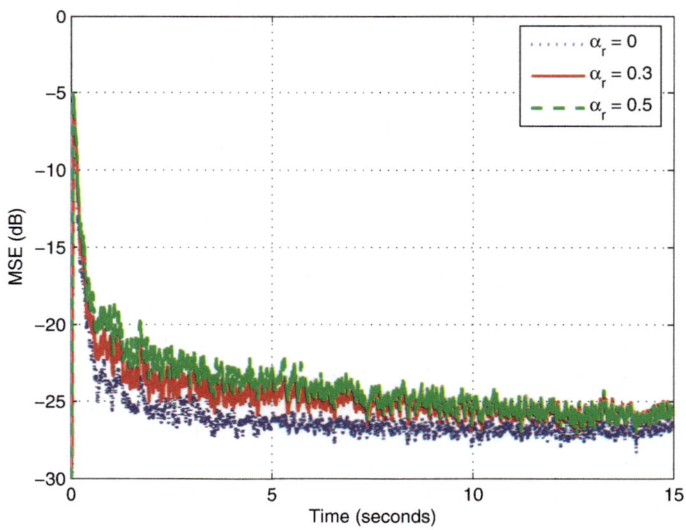

Fig. 9.39 MSE of the APA with $P = 8$. Other conditions same as in Fig. 9.38.

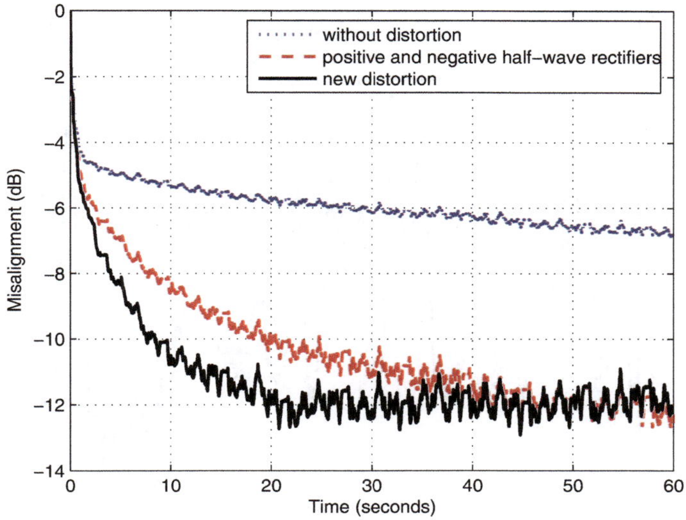

Fig. 9.40 Misalignment of the APA with $P = 8$ for different types of distortion with $\alpha_r = 0.3$. The source signal is a speech sequence.

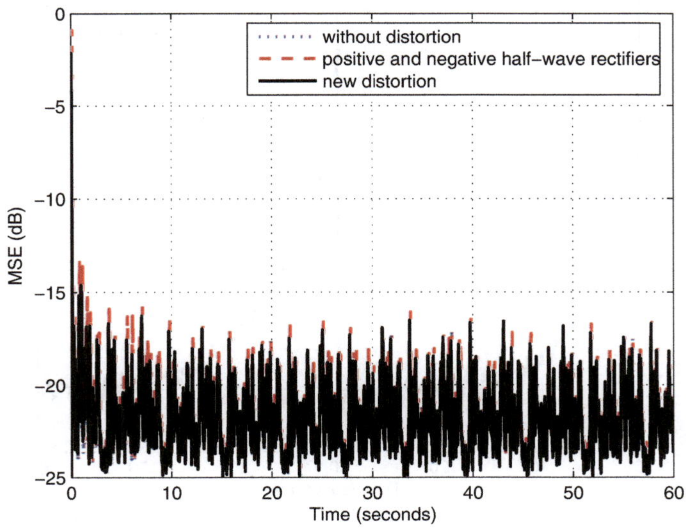

Fig. 9.41 MSE of the APA with $P = 8$ for different types of distortion. Other conditions same as in Fig. 9.40.

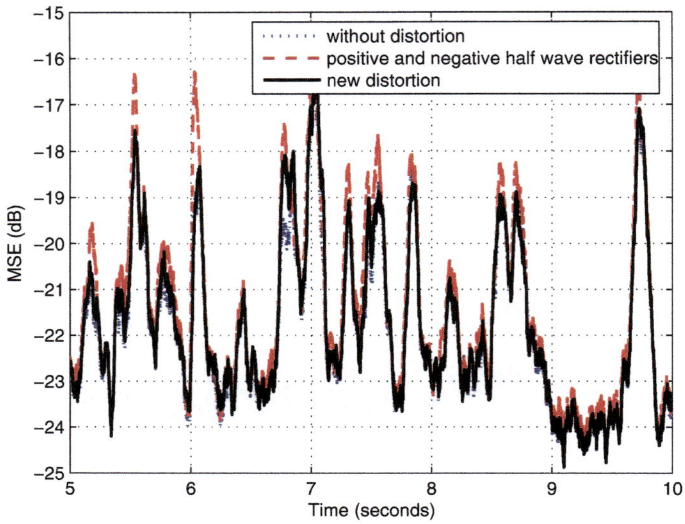

Fig. 9.42 MSE of the APA with $P = 8$ for different types of distortion. Detail of Fig. 9.41.

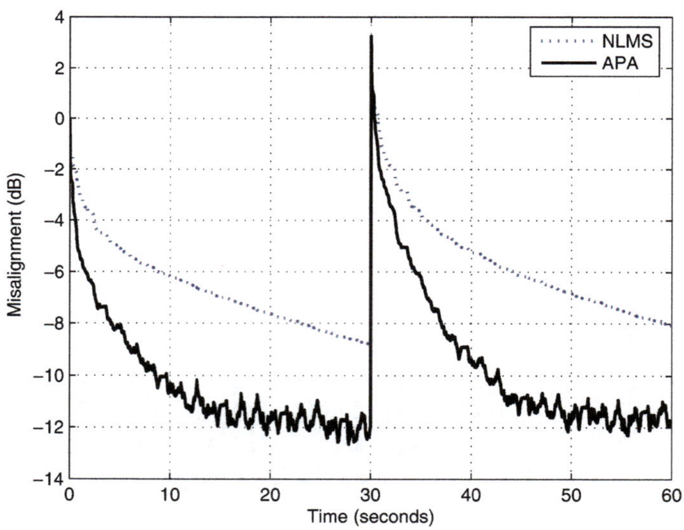

Fig. 9.43 Misalignment of the NLMS algorithm and the APA with $P = 8$ in a tracking situation. The source signal is a speech sequence and the new distortion is used with $\alpha_r = 0.3$.

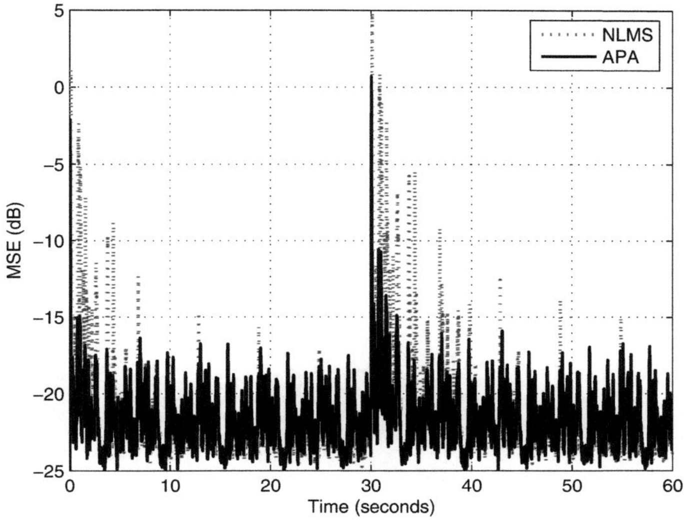

Fig. 9.44 MSE of the NLMS algorithm and the APA. Other conditions same as in Fig. 9.43.

fact, this parameter is estimated within the algorithm, but using only parameters that are available from the adaptive filter and the observation signal, $d(n)$. Figure 9.45 compares the misalignment of the APA using $\alpha = 0.25$ with the misalignment of the VSS-APA, while the corresponding MSE curves are given in Fig. 9.46; the projection order is $P = 8$. The source signal is white Gaussian and the far-end microphone signals are distorted using positive and negative half-wave rectifiers with $\alpha_{\mathrm{r}} = 0.3$. It can be noticed that the VSS-APA converges slightly faster than the fixed step-size APA but achieves a lower MSE level.

The idea of the IPNLMS algorithm [6] was straightforwardly extended to the APA, resulting the IPAPA [9]. This algorithm is presented in Section 5.5, Chapter 5. First, we evaluate its capabilities as compared to the IPNLMS algorithm. Figure 9.47 compares the misalignment of the IPAPA using different projection orders (i.e., $P = 2$, 8, or 16) with the misalignment of the IPNLMS algorithm (which is equivalent to the IPAPA with $P = 1$). The corresponding MSE curves are plotted in Fig. 9.48. The source signal is white Gaussian and the positive and negative half-wave rectifiers with $\alpha_{\mathrm{r}} = 0.3$ are used to distort the far-end microphone signals. The normalized step size for all the algorithms is set to $\alpha = 0.25$, the regularization parameter is $\delta = 20\sigma_x^2/(2L)$, and the proportionate parameter is $\kappa = 0$. If not specified otherwise, these values will be used in all the following experiments involving the IPAPA. As expected, the convergence rate of the IPAPA increases when the value of the

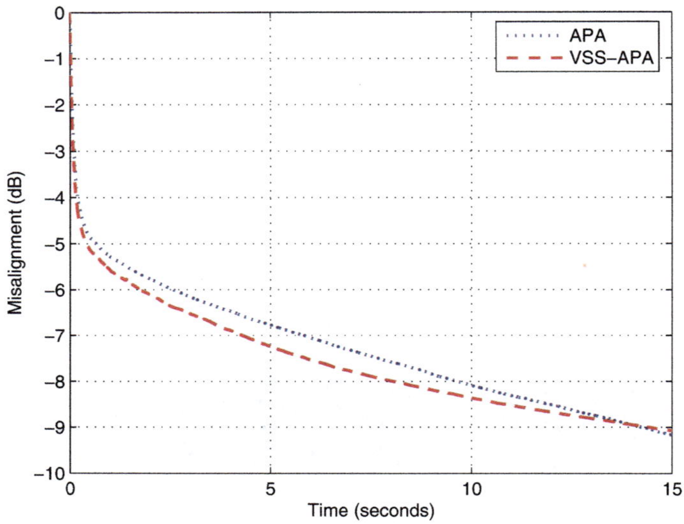

Fig. 9.45 Misalignment of the APA (with $\alpha = 0.25$) and the VSS-APA, with $P = 8$. The source signal is white Gaussian. Preprocessing with positive and negative half-wave rectifiers and $\alpha_{\mathrm{r}} = 0.3$.

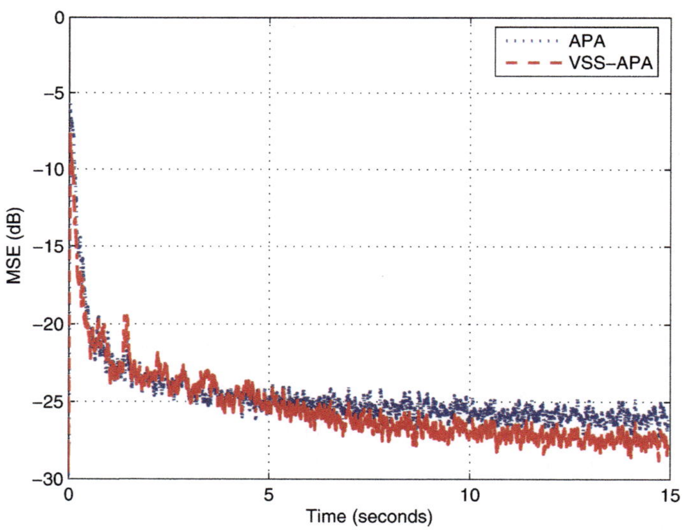

Fig. 9.46 MSE of the APA and the VSS-APA. Other conditions same as in Fig. 9.45.

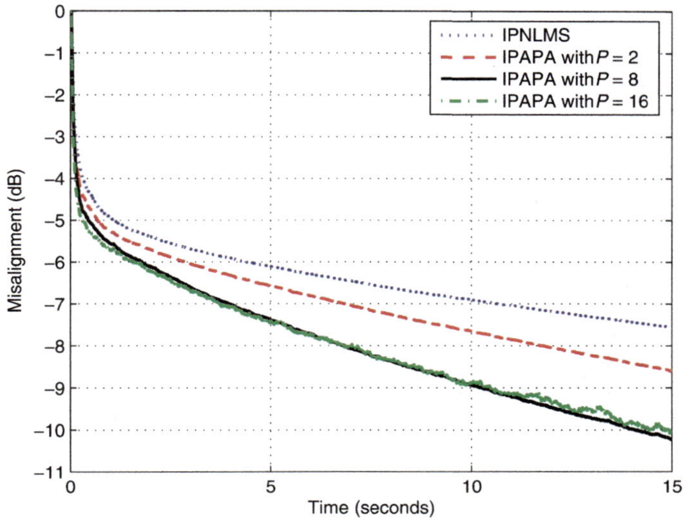

Fig. 9.47 Misalignment of the IPNLMS algorithm and the IPAPA using different values of the projection order. The source signal is white Gaussian. Preprocessing with positive and negative half-wave rectifiers and $\alpha_{\mathrm{r}} = 0.3$.

projection order increases; however, it is not worth to use a projection order higher than $P = 8$. Also, as we can notice from Fig. 9.48, the MSE of the IPAPA also increases with the projection order. Since the value $P = 8$ offers a proper compromise, this value of the projection order will be used in all the following experiments involving the IPAPA.

Next, we evaluate the impact of the proportionate parameter κ over the performance of the IPAPA. Figure 9.49 shows the misalignment of the IPAPA using different values of the parameter κ, as compared to the APA. The corresponding MSE curves are provided in Fig. 9.50. The source signal is white Gaussian and the far-end microphone signals are distorted using positive and negative half-wave rectifiers with $\alpha_{\mathrm{r}} = 0.3$. Figure 9.49 justifies that the value $\kappa = 0$ represents a proper choice for the proportionate amount; this value will be used in all the following experiments involving the IPAPA. According to Fig. 9.50, all the algorithms perform very similarly in terms of the MSE.

Figure 9.51 compares the misalignment of the IPAPA using positive and negative half-wave rectifiers versus the new distortion proposed in Chapter 3 [see (3.38) and (3.39)]; also, the case without distortion is shown as a reference. The input source is a speech sequence and the distortion parameter is set to $\alpha_{\mathrm{r}} = 0.3$. The corresponding MSE curves are depicted in Fig. 9.52. Similar to the APA, it can be noticed from Fig. 9.51 that the IPAPA converges faster with the new distortion. Besides, as we can see in Fig. 9.52 and in the detail presented in Fig. 9.53, the new distortion leads to a slightly

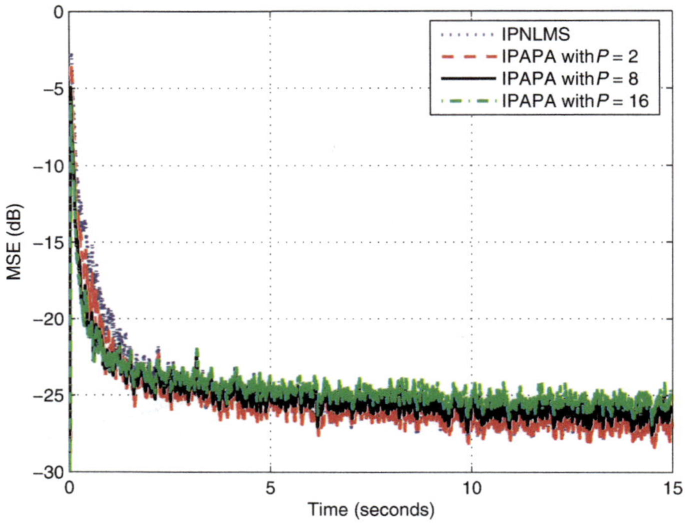

Fig. 9.48 MSE of the IPNLMS algorithm and the IPAPA using different values of the projection order. Other conditions same as in Fig. 9.47.

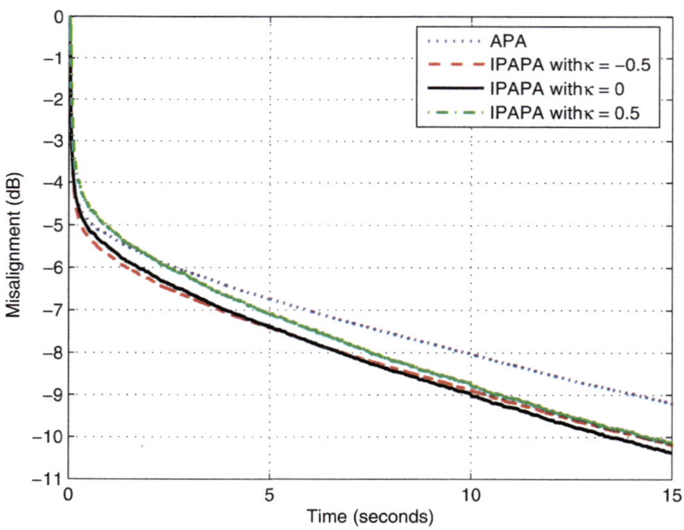

Fig. 9.49 Misalignment of the APA and the IPAPA using different values of the parameter κ; the projection order is $P = 8$. The source signal is white Gaussian. Preprocessing with positive and negative half-wave rectifiers and $\alpha_{\rm r} = 0.3$.

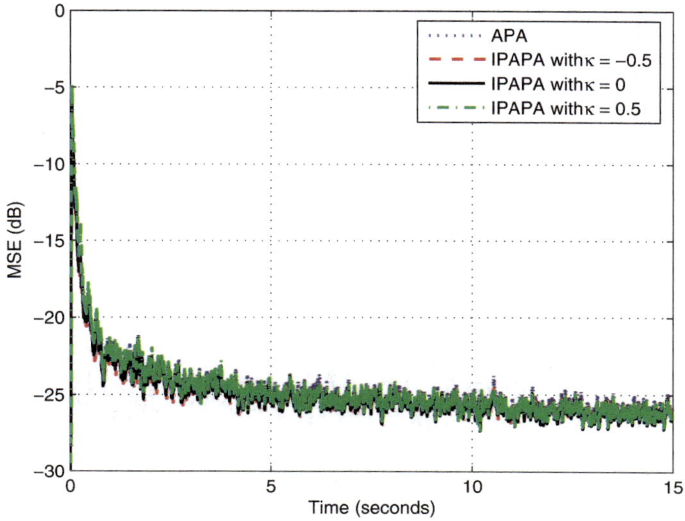

Fig. 9.50 MSE of the APA and the IPAPA using different values of the parameter κ. Other conditions same as in Fig. 9.49.

better performance in terms of the MSE as compared to the positive and negative half-wave rectifiers.

Since the IPAPA has resulted as a combination between the IPNLMS algorithm and the APA, it is expected that the IPAPA should outperform both its predecessors. The following experiment outlines this aspect, by comparing these three algorithms in a tracking situation. The source signal is speech and the new distortion is used with $\alpha_\mathrm{r} = 0.3$. All the algorithms use the same normalized step size $\alpha = 0.25$, the IPNLMS and IPAPA use $\kappa = 0$, and $P = 8$ for the APA and IPAPA. The results are shown in Figs. 9.54 (for the misalignment) and 9.55 (for the MSE). According to these plots, it is clear that IPAPA outperforms both the IPNLMS and APA.

The MPAPA presented in Section 5.6 takes advantage of the "proportionate memory," by taking into account the "history" of the proportionate factors from the last P steps. This specific feature of the MPAPA leads to efficient recursive implementations of its parameters. Therefore, the MPAPA is more computationally efficient as compared to the classical PAPAs. The recently proposed MIPAPA [11] has resulted as a combination between the idea of the MPAPA and the proportionate factors of the IPAPA.

In the following experiment, the MIPAPA is compared to the IPAPA in a tracking situation. The proportionate parameter for both algorithms is $\kappa = 0$, the projection order is $P = 8$, the normalized step size is set to $\alpha = 0.25$, and the regularization parameter is $\delta = 20\sigma_x^2/(2L)$. The input source is

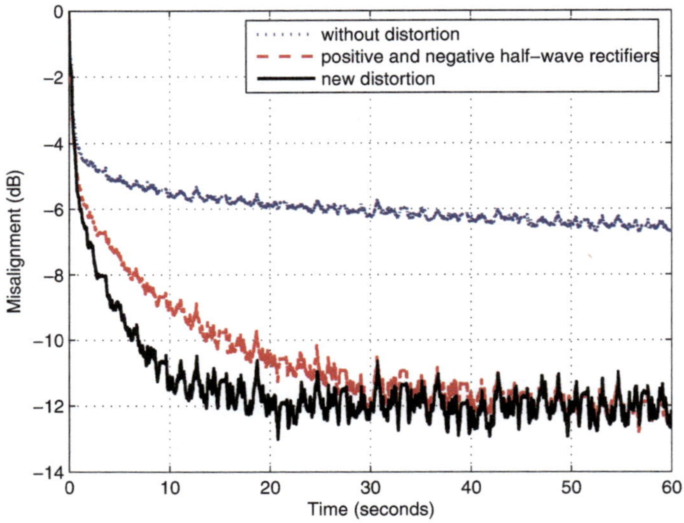

Fig. 9.51 Misalignment of the IPAPA with $P = 8$ and $\kappa = 0$ for different types of distortion with $\alpha_r = 0.3$. The source signal is a speech sequence.

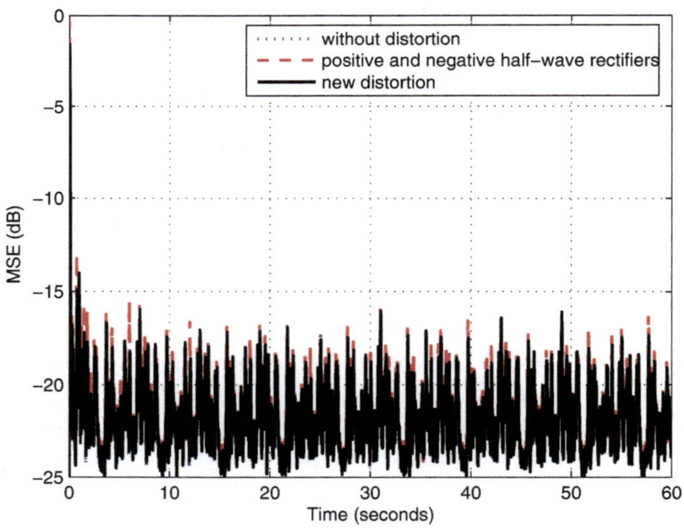

Fig. 9.52 MSE of the IPAPA with $P = 8$ and $\kappa = 0$ for different types of distortion. Other conditions same as in Fig. 9.51.

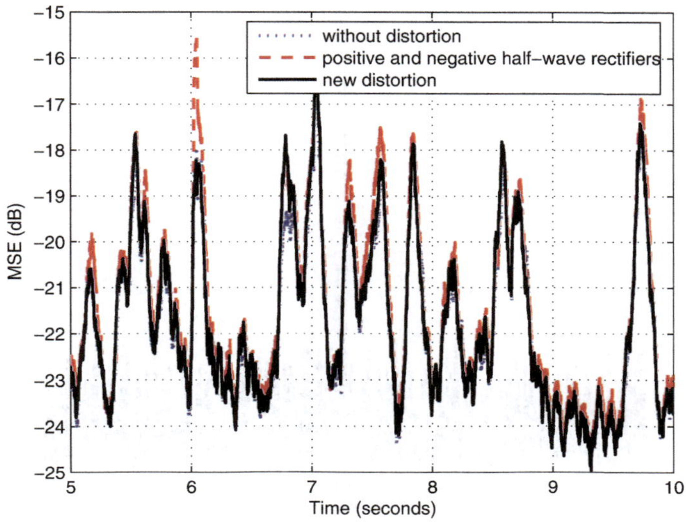

Fig. 9.53 MSE of the IPAPA with $P = 8$ and $\kappa = 0$ for different types of distortion. Detail of Fig. 9.52.

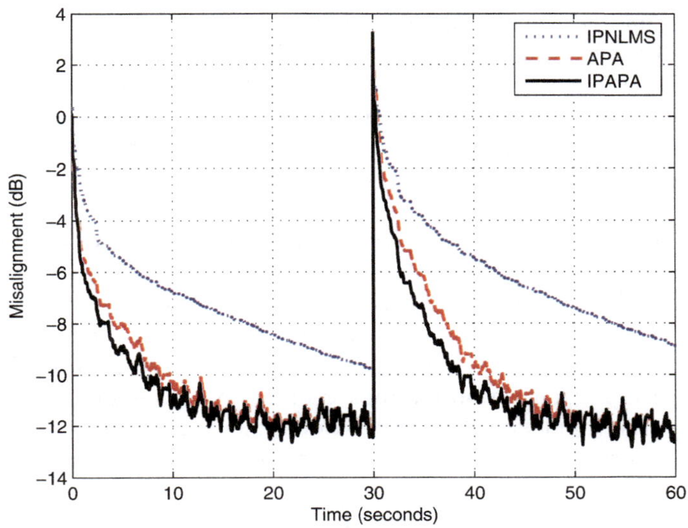

Fig. 9.54 Misalignment of the IPNLMS algorithm, the APA, and the IPAPA in a tracking situation; $\kappa = 0$ and $P = 8$. The source signal is a speech sequence and the new distortion is used with $\alpha_r = 0.3$.

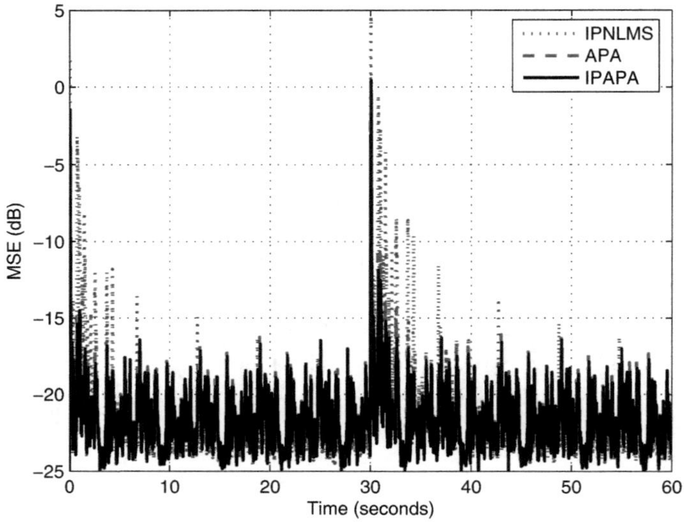

Fig. 9.55 MSE of the IPNLMS algorithm, the APA, and the IPAPA in a tracking situation. Other conditions same as in Fig. 9.54.

white Gaussian and the microphone signals are distorted using positive and negative half-wave rectifiers with $\alpha_r = 0.3$. Figure 9.56 compares the IPAPA and the MIPAPA in terms of the misalignment, while the associated MSE curves are depicted in Fig. 9.57. It can be noticed that the MIPAPA slightly outperforms the IPAPA in terms of tracking. However, we should remember that the computational complexity of the MIPAPA is lower as compared to the IPAPA.

Regularization also plays an important role within the APAs, especially for low SENRs. In Section 5.3, it was derived an optimal regularization parameter for the APA. The optimal normalized regularization parameter of the APA, denoted by β_{APA}, is given in (5.40). It can be noticed that the regularization parameter of the APA does not depend on the projection order P and is identical to the regularization parameter of the NLMS algorithm when we assume that the input signal is white. Consequently, similar to the case of the NLMS algorithm, the importance of β_{APA} becomes more apparent for low SENRs.

Figure 9.58 compares the misalignment of the APA (with $P = 8$) using the optimal β_{APA} with the ad-hoc choice $\beta = 20$, when the SENR is set to 10 dB and when a tracking situation is considered. The corresponding MSE curves are provided in Fig. 9.59. The source signal is white Gaussian and the microphone signals are distorted using positive and negative half-wave rectifiers with $\alpha_r = 0.3$. According to these results, the APA using the

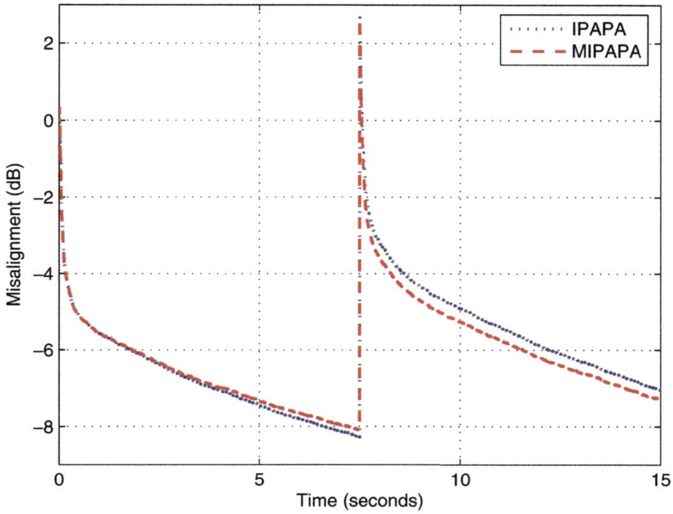

Fig. 9.56 Misalignment of the IPAPA and the MIPAPA, with $P = 8$ and $\kappa = 0$. The source signal is white Gaussian. Preprocessing with positive and negative half-wave rectifiers and $\alpha_{\rm r} = 0.3$.

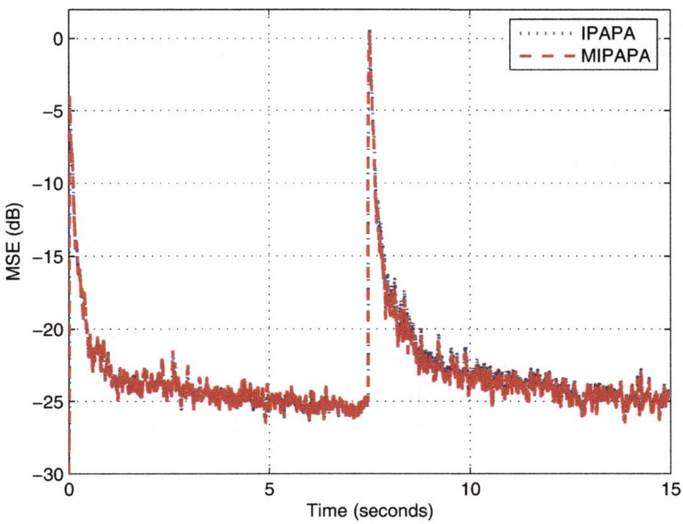

Fig. 9.57 MSE of the IPAPA and the MIPAPA. Other conditions same as in Fig. 9.56.

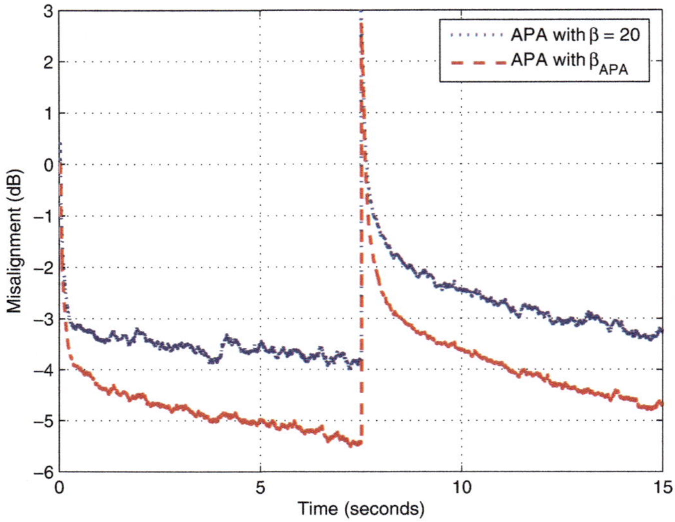

Fig. 9.58 Misalignment of the APA (with $P = 8$) using $\beta = 20$ and β_{APA}, when SENR = 10 dB; a tracking situation is considered. The source signal is white Gaussian. Preprocessing with positive and negative half-wave rectifiers and $\alpha_{\mathrm{r}} = 0.3$.

optimal regularization outperforms the classical regularization, in terms of both the misalignment and MSE.

9.4 FRLS Algorithm

RLS-type algorithms represent a very attractive choice in many applications, mainly due to their fast convergence rate. However, the classical RLS algorithm is not a good choice for SAEC due to its computational complexity. Taking into account the computational issues, the FRLS algorithm is a practical alternative to the RLS. Nevertheless, the FRLS algorithm is not always easy to control in practice in terms of its numerical stability. In this section, we briefly outline the capabilities of the FRLS algorithm as compared to the classical benchmarks, i.e., NLMS and APA.

It is very important to correctly choose the main parameters of the FRLS algorithm, i.e., the forgetting factor λ_{L} and the initialization parameters $\mathbf{E}_{\mathrm{f}}(0)$ and $\mathbf{E}_{\mathrm{b}}(0)$ (see Section 6.3 in Chapter 6); otherwise, the algorithm could become unstable. In our simulations, we set $\lambda_{\mathrm{L}} = 1 - 1/(12L)$.

In the following experiment the source signal is white Gaussian and the positive and negative half-wave rectifiers (with $\alpha_{\mathrm{r}} = 0.3$) are used to distort the far-end microphone signals. The FRLS algorithm is compared with the

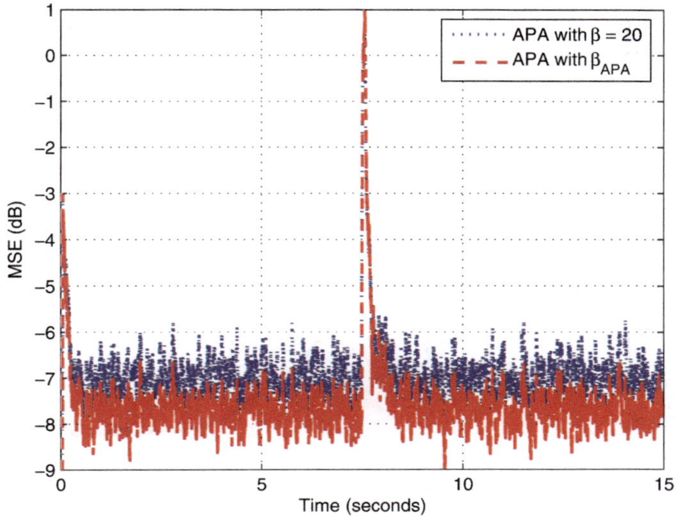

Fig. 9.59 MSE of the APA using $\beta = 20$ and β_{NLMS}. Other conditions same as in Fig. 9.58.

NLMS and APA, both using the normalized step size $\alpha = 0.25$ and the regularization parameter $\delta = 20\sigma_x^2$. The projection order for the APA is $P = 8$. The results are shown in Figs. 9.60 (for the misalignment) and 9.61 (for the MSE). In this simulation example, even if the APA is slightly superior in terms of the initial convergence rate, it is clear that the FRLS algorithm outperforms the other algorithms in terms of both the misalignment and MSE.

References

1. J. Benesty, T. Gänsler, D. R. Morgan, M. M. Sondhi, and S. L. Gay, *Advances in Network and Acoustic Echo Cancellation.* Berlin, Germany: Springer-Verlag, 2001.
2. J. Benesty, D. R. Morgan, and M. M. Sondhi, "A better understanding and an improved solution to the specific problems of stereophonic acoustic echo cancellation," *IEEE Trans. Speech, Audio Process.*, vol. 6, pp. 156–165, Mar. 1998.
3. J. Benesty, H. Rey, L. Rey Vega, and S. Tressens, "A non-parametric VSS-NLMS algorithm," *IEEE Signal Process. Lett.*, vol. 13, pp. 581–584, Oct. 2006.
4. D. L. Duttweiler, "Proportionate normalized least-mean-squares adaptation in echo cancelers," *IEEE Trans. Speech, Audio Process.*, vol. 8, pp. 508–518, Sept. 2000.
5. C. Paleologu, J. Benesty, and S. Ciochină, *Sparse Adaptive Filters for Echo Cancellation.* San Rafael: Morgan & Claypool, 2010.
6. J. Benesty and S. L. Gay, "An improved PNLMS algorithm," in *Proc. IEEE ICASSP*, 2002, pp. 1881–1884.

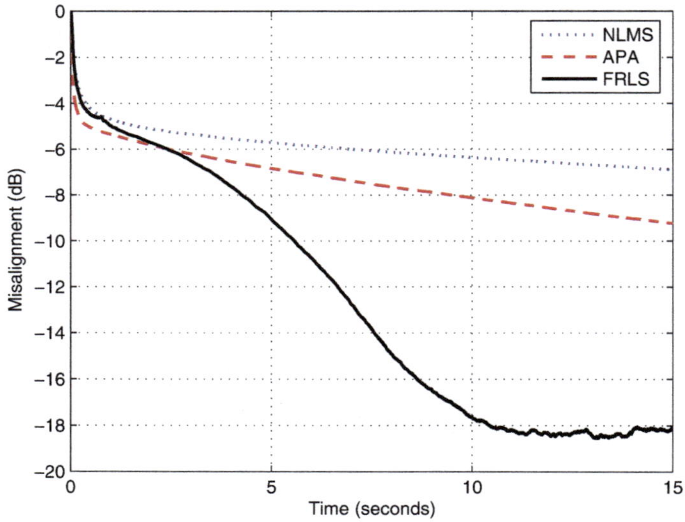

Fig. 9.60 Misalignment of the NLMS, AP, and FRLS algorithms. Parameters: $\alpha = 0.25$, $P = 8$, and $\lambda_{\mathrm{L}} = 1 - 1/(12L)$. The source signal is white Gaussian. Preprocessing with positive and negative half-wave rectifiers and $\alpha_{\mathrm{r}} = 0.3$.

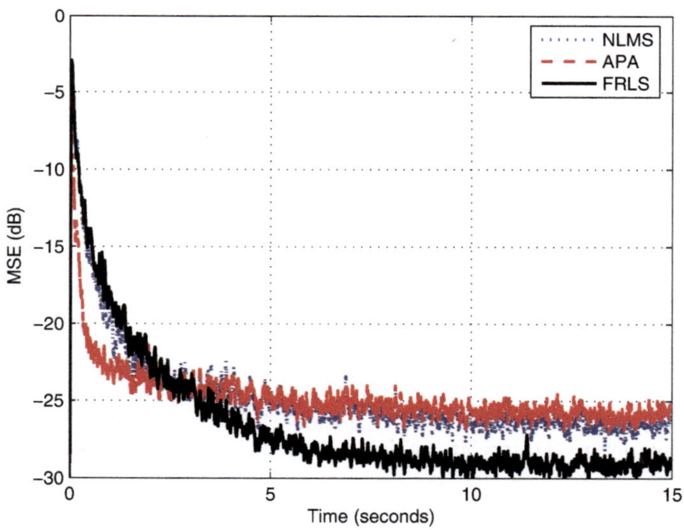

Fig. 9.61 MSE of the NLMS, AP, and FRLS algorithms. Other conditions same as in Fig. 9.60.

7. K. Ozeki and T. Umeda, "An adaptive filtering algorithm using an orthogonal projection to an affine subspace and its properties," *Electron. Commun. Jpn.*, vol. 67-A, pp. 19–27, May 1984.
8. C. Paleologu, J. Benesty, and S. Ciochină, "A variable step-size affine projection algorithm designed for acoustic echo cancellation," *IEEE Trans. Audio, Speech, Language Process.*, vol. 16, pp. 1466–1478, Nov. 2008.
9. O. Hoshuyama, R. A. Goubran, and A. Sugiyama, "A generalized proportionate variable step-size algorithm for fast changing acoustic environments," in *Proc. IEEE ICASSP*, 2004, pp. IV-161–IV-164.

Index

GPSR Compliance

The European Union's (EU) General Product Safety Regulation (GPSR) is a set of rules that requires consumer products to be safe and our obligations to ensure this.

If you have any concerns about our products, you can contact us on ProductSafety@springernature.com

In case Publisher is established outside the EU, the EU authorized representative is:

Springer Nature Customer Service Center GmbH
Europaplatz 3
69115 Heidelberg, Germany

Batch number: 10614746

Printed by Printforce, the Netherlands